THE SHOEMAKER'S HOLIDAY

A section of Claes Jans Visscher's 'Long View of London', 1616, showing Leadenhall

THOMAS DEKKER

THE
SHOEMAKER'S
HOLIDAY

EDITED BY J.B.STEANE

CAMBRIDGE
AT THE UNIVERSITY PRESS
1965

PUBLISHED BY

THE SYNDICS OF THE CAMBRIDGE UNIVERSITY PRESS

Bentley House, 200 Euston Road, London, N.W. 1
American Branch: 32 East 57th Street, New York, N.Y. 10022
West African Office: P.O. Box 33, Ibadan, Nigeria

CAMBRIDGE UNIVERSITY PRESS
1965

Printed in Great Britain at the University Printing House, Cambridge
(Brooke Crutchley, University Printer)

LIBRARY OF CONGRESS CATALOGUE
CARD NUMBER: 65-15305

CONTENTS

INTRODUCTION

I

In some ways this is the most approachable of Elizabethan plays. It is quite possible for an intelligent person to go to a performance of *King Lear* and not make head nor tail of what Edgar is doing or exactly how events take shape in the last act. A television production of *The Alchemist* lost half its audience in the first fifteen minutes because the language was so unfamiliar and the pace of speech so fast that people were left behind and gave up the struggle before the action had really begun. This could not happen in *The Shoemaker's Holiday*. Its language is clear and its story easy to follow. Yet there *is* a barrier between us and Dekker's play, and it goes rather deeper than whatever it was that put the television audience off Ben Jonson. For, after all, the humour of *The Alchemist* is remarkably modern: quick-witted, likeable crooks make their fortune and their fun out of fools, but are eventually discomfited themselves. The humour of discomfiture appeals to our age as it did to the Elizabethans. So does the smart, wise-cracking dialogue that is contrasted with the ludicrous, stylised speech of the eccentrics whom the crooks deceive. But *The Shoemaker's Holiday* offers a kindly humour, or rather—and here comes the deadly term—a hearty humour. A 'jocular, back-slapping patriotic piece' it is called by the writer in the Pelican Guide,[1] and there could hardly be a description less designed to whet the appetite of most modern readers, old or young. For they are unlikely to go to literature as a kind of substitute for a scout camp-fire.

[1] Pelican Guide to English Literature. Vol. 2. The Age of Shakespeare: D. J. Enright, *Elizabethan and Jacobean Comedy*, p. 419.

From anything back-slapping we tend to flinch; at anything patriotic we tend to smile. And 'hearty' is nowadays a term of abuse; almost as automatic and derogatory as 'suburban', 'middle-class' or the other terrible epithets with which people damn the things they dislike.

But if our sophistication prevents enjoyment of such a comedy as this, the loss is ours. Dekker provides an entertainment; tells a tale, creates characters, hopes to infect an audience with the good-humoured zest of contented workers, ardent lovers, banquets and holidays. He never mocks, rarely satirises. But then, ours is the age which has rediscovered satire: that is the kind of humour to which we have most conspicuously taken. Other kinds of humour still exist: the fantasy of Ionesco, the inconsequential humour of Beckett or Pinter, the irreverent sniping of Kingsley Amis's Lucky Jim, or the sentimental humour of the little man (H. G. Wells and Chaplin providing the prototypes, Thurber's Walter Mitty the archetype). The simple humour of high spirits is not in fashion. But this is what Dekker has to give: innocence (in spite of the bawdry), freshness and energy—characteristics that are valuable in him and the literature of his age in proportion to their rarity in ours.

II

He takes us, for most of his scenes, to a London very different from that of the present day. We are much aware of the City, through the frequent mention of place-names (Guildhall, Leadenhall, St Paul's, the Savoy, the Conduits, London Stone), through the sense of municipal dignity in the Lord Mayor and Simon, and through the bustle of a populous place with its business transactions and vigorous tradesmen. The Dutch ship which has come in laden with cloths, sugar and spices all going at bargain price is a kind of Elizabethan dream; such good fortune cannot have presented itself often, yet it gives us a picture of the commerce through which middle-class men could become wealthy and influential. With riches would come honour and authority. This meant an important position in one of the trade guilds, then perhaps the rank of Alderman, and possibly even that of Lord Mayor. Such is Simon Eyre's story and as we follow his career it is always with a consciousness of the world in which he rises, the London of great Companies, of masters and prentices, working days and festivals.

Dekker was a great Londoner. London, he says, was his cradle: 'from thy womb received I my being, from thy breasts my nourishment'.[1] His feeling for the City was strong throughout his life and is reflected in the liveliness of his observation. He brings the sights and sounds vividly before us in this passage from *The Seven Deadly Sins of London* (1606):

In every street, carts and coaches make such a thundering as if the world ran upon wheels; at every corner men, women and children meet in such shoals that posts are set up of purpose to strengthen the houses, lest with justling one another they [the people] should shoulder them down. Besides, hammers are beating in one place, tubs hooping in another, pots clinking in a third, water tankards running at tilt in a fourth: here are porters sweating under burdens, there merchants' men bearing bags of money, chapmen (as if they were at leap-frog) skip out of one shop into another, tradesmen (as if they were dancing galliards) are lusty at legs and never stand still. All are as busy as country attornies at an Assises. How then can Idleness think to inhabit here?

This is very much the world of *The Shoemaker's Holiday*, where Eyre and his prentices scurry round from first thing in

[1] *The Seven Deadly Sins of London* (The Non-Dramatic Works of Thomas Dekker, ed. Grosart, vol. 2).

the morning, and where the words, the banter and cross-talk, come tumbling out at a speed to match. It is the vigorous London that contained Shakespeare's Eastcheap and Jonson's Bartholomew Fair; and in Dekker, probably above all others, we feel a closeness to the day-to-day world of ordinary folk.

Along with the bustle, there was also dignity, and in spite of the congestion a certain spaciousness. The river was still London's main thoroughfare. From the Queen in her royal barge to the folk who paid twopence for the 'common barge' all sorts and conditions would travel by water. They would then see the City much as it appears in the old views of London (see the reproduction of Visscher's 'Long View' on the frontispiece), a place of fine buildings overlooked by 120 church spires. As Lord Mayor, Simon would have had his own state barge, and as a member of the Company of Shoemakers (Cordwainers) he would probably have used theirs also. The Companies and officials did much to preserve the solid dignity of the town, the sense of its traditions, and the colour of its ceremonies. This too is reflected in the play.

Pride in King, country and city rings throughout *The Shoemaker's Holiday*, and is linked with the pride in a great city craft, conscious of its identity and standing. Like other trades in the Middle Ages, the 'gentle craft' of shoemaking was closely organised and controlled. Simon Eyre, not just an invention of Dekker's (see pp. 137–40), but in historical fact Lord Mayor of London in 1436, lived at a time when the power of the Guilds was at its height. Early in the Middle Ages the trades had formed themselves into fraternities with a religious basis and an economic function. Each guild furthered the interests of its craft by mutual protection, and in the early fourteenth century they received formal recognition by the granting of charters from the King. They became known as companies: 'livery companies' in London

4

because the members wore a costume or livery distinctive to their particular trade. There was now a well-defined hierarchy in the trades, a kind of human pyramid with a multitude of apprentices at the base and a Master of the Company on high. An apprentice was bound to the service of a master, learning under conditions very specifically laid down by the guild, for a period of indenture nearly always lasting seven years. He would then become a journeyman; that is, a skilled craftsman working for a daily wage and free to contract his services to any master (hence often itinerant, or 'journeying'—Hodge and Firke are in this position at the beginning of the play and so they are able to threaten strike action, as in an elementary way they do, twice). In the play we later see the senior journeyman, Hodge, become master himself, Simon Eyre having meantime been promoted to higher things. A master might well belong to the City Company, which in turn had its own hierarchy. The freemen of the Company were recruited partly on a hereditary basis, partly by nominal apprenticeship to a member. From the freemen the 'livery' was recruited: that is, a body of senior members who were likely to be chosen for the highest offices. When so chosen they would become members of the Court, the governing body of the Company over which the Master presided.

These men were prosperous and prominent citizens and their authority did not end within the Company, for the City looked to them for its Aldermen and for its Mayor. It was no doubt such social advantages (bringing with them economic profit too) that guaranteed the continuance of the Companies to the present day, long after their effectiveness as trade organisations has declined. Most of them (including the Cordwainers) are in existence today, principally as administrators of trusts and charities and as property owners;

and it is still the freemen of the liveried companies who elect the Lord Mayor.

The Company of Cordwainers dates from the thirteenth century, its name deriving from Cordova (Corduba), the Spanish town which produced the goatskin from which the finest boots and shoes were made. Much of the Cordwainers' energy seems to have been devoted to skirmishes with the cobblers, who were looked on as inferiors, as shopkeepers rather than craftsmen, and who moreover sold old shoes which they had repaired (a flourishing repair trade was obviously not good for the shoemakers' business). In Dekker's play we hear Eyre refer to 'the gentlemen shoemakers, the courageous cordwainers' (I, 1, 224–5), but the word 'cobbler' never passes his lips or anyone else's. This was evidently a craft very much aware of its rights and its dignity. Not one of the twelve great City Companies,[1] the Cordwainers were, nevertheless, fairly well-off compared with the sixty-or-so minor companies that were in existence, and Eyre's period was a particularly good one for the shoemakers. The play ends with him as Lord Mayor and in great prosperity, having successfully petitioned the King (Henry VI) for the conferring of privileges upon his craft (see v, 5, 158–68).[2] The jollifications of the play all have sound economic justification, and in the joviality of Dekker's entertainment we also gain a vivid and valuable glimpse into City life as it was in his time and as it had been for several centuries before him.

[1] These were: the Mercers, Grocers, Drapers, Fishmongers, Goldsmiths, Skinners, Merchant Taylors, Haberdashers, Salters, Ironmongers, Vintners and Clothworkers. In *The Dead Term* (1608) Dekker described London as 'The Mother of the twelve Companies'.

[2] Henry VI conferred a charter upon the freemen of the Mistery of the Cordwainers in 1439, giving them powers of search and control over all workers in black or red tanned leather and new boots and shoes within the City of London and a radius of two miles from Cornhill (where Simon built his Hall).

For this picture of the workings of a trade, with its own practices and its connection with municipal government, reflects the essential organisation of the towns, as the feudal system embodied that of the country.

Moreover, *The Shoemaker's Holiday* takes us close to the ordinary people of the town in a way that is not common. Simon and Mistress Eyre with their hearty bourgeois liveliness represent a part of London that we meet again in Beaumont and Fletcher's *Knight of the Burning Pestle* but not much elsewhere. For in Elizabethan drama the heroic, 'serious' characters are generally aristocratic, while the lower classes are merely comic or incidental. In this play we are more faithfully introduced to the centre of London life than we are by Shakespeare, or even by the Bartholomew birds of Jonsonian comedy.

III

This centrality should not be surprising, for Dekker strikes one in most of his writing as a sensible observer, whose feelings and interests correspond reliably with those of the sound, level-headed majority. He is not a profound or original thinker, but that very limitation increases his 'documentary' value. 'He does not draw on popular thought and refine it, like Jonson', says L. C. Knights, 'his thoughts *are* the thoughts of the average Londoner.'[1]

As a social commentator, however, he is very incompletely represented by *The Shoemaker's Holiday*, and it will be necessary if we are to have an understanding of our author to look briefly into some of his other works, particularly the prose pamphlets. In *The Shoemaker's Holiday* the mood is gay, very much as caught in the two 'Three-mans songs'.

[1] *Drama and Society in the Age of Jonson* (first published 1937; Peregrine Books, 1962), see ch. VIII.

Just occasionally there are darker patches: in Ralph's war-wounds, perhaps, or Jane's grief and perplexity, and the efforts of Mayor and nobleman to thwart the course of true love. But none of these is allowed to shut out the sunshine for very long, and the prevailing tone is set by the high spirits of Eyre and his men. 'A merry cobbler there was, who for joy that he mended men's broken and corrupted soles did continually sing, so that his shop seemed a very bird-cage.'[1] Dekker wrote this in another context, but the description (barring the term 'cobbler' of course) suits his own shoe-makers well and suggests the spirit of the play. It is not the spirit of many of his later works, however.

He had, for instance, an intense feeling for the wrongs suffered by the poor. On the title-page of *Work for Armourers* (1609)[2] he wrote the motto: 'God help the poor, the Rich can shift.' The pamphlet contains much sharp analysis of the class struggle and its last words are as biting as its first: 'The rich men feast one another, as they were wont, and the poor were kept poor still, in policy, because they should do no more hurt.' With a similar bitterness several other works set about attacking the 'army of insufferable abuses' that Dekker saw around him. *The Seven Deadly Sins of London* is well worth turning up both as an example of his thought and style, and as a social document. Among the deadly seven is one surprisingly called 'shaving'. This turns out to be what communists might describe as 'capitalist blood-sucking' or hard, extortionate exploitation of the poor. 'Cruel and covetous landlords', he speaks of, 'who for the building up of a chimney, which stands them not above 30s. and for whiting the walls of a tenement, which is scarce worth the daubing, raise the rent presently... assessing it three pounds a year more than ever it went for before'. The seventh and

[1] *The Raven's Almanac*, Grosart, vol. 4 (1609).   [2] Grosart, vol. 4.

worst of the sins is cruelty. Perhaps the most bitter passage here is this:

Look again over thy walls into thy fields, and thou shalt hear the poor and forsaken wretches lie groaning in ditches, and travailing to seek out Death upon thy common highways. Having found him, he there throws down their infected carcasses, towards which all that pass by look, but (till common shame and common necessity compel) none step in to give them burial. Thou setst up posts to whip them when they are alive—set up an hospital to comfort them being sick! Or purchase ground for them to dwell in when they be well, and that is, when they be dead.

The 'thou' being addressed is the respectable, decent, well-to-do Londoner; and the poor particularly in mind here are the army of beggars created by enclosures, tormented by the plague, and about to break out into riots, all ruthlessly suppressed, in 1607. Dekker's sympathy is clear; so is the social indignation that burns along with the compassion.

One would very much like that to be a complete account of Dekker's attitude on these problems. It would then be possible to point to him as 'the best democrat of his age' as some have done[1] and leave it at that. But there is a strongly conservative side to him, quite as genuine as the other, and rather hard to reconcile with it. Dr K. L. Gregg[2] discusses this and quotes from *Four Birds of Noah's Ark* (1609):[3]

The maid servant prays that, as the Lord has laid upon her the condition of a servant, her mind may be subjected to the state in which she was placed, and the serving man consoles himself in the thought that in the service of the Lord, he has a promotion greater than that due to kings.

This is the perfectly traditional acceptance of that condition 'to which the Lord has called us'. The serenity of this, and

[1] A. F. Lange quoted in *Thomas Dekker: a Study in Economic and Social Backgrounds* by K. L. Gregg (University of Washington Publications, Language and Literature, II, ii, pp. 52–112, July 1924).
[2] *Op. cit.*   [3] Grosart, vol. 5.

much else in Dekker,[1] is certainly in key with the mind that created *The Shoemaker's Holiday*, and it is certainly not the mind of a ruthless root-and-branch social critic. Yet that side of Dekker does exist also and it is interesting to inquire how the two could live together in the same man (for they continued to do so, even though the bitterness seems to have grown in later years). Dr Gregg concludes that 'admiration for the theory of the past and contempt for the manner in which it works' is 'typically English' (she is American); so perhaps even in this apparent contradiction within Dekker as a social writer we have another example of his centrality, thinking 'the thoughts of the average Londoner'.

IV

That Dekker's mind did grow more critical as time went on is evident even from the matter of masters and apprentices as dramatised in *The Shoemaker's Holiday*. In this, all is well; Eyre is a good master, his journeymen are good lads. There is a hint of early trade-unionism in II, 3, where Firke takes umbrage, lays down tools and prepares to leave, only to be followed, with a splendid show of class-solidarity, by Hodge: 'Nay, stay, Firke. Thou shalt not go alone.'[2] But it is all part of the comedy, and 'a dozen cans of beer' (or, in reality, two) from 'The Boar's Head' puts all right again. There is never any real discontent or any serious rub of class-feeling. That was in 1599. In 1606, the year of *The Seven Deadly Sins of London*, we find him writing of the relationship between

[1] See also the play *Patient Grissil* (1599) where every kind of monstrous trial seems to be justified on the grounds that it tests a loyal and obedient spirit. There is also an unquestioning conservatism in the speech of the father (Ianicola) who upbraids his scholar son for trying to 'kick against the faults of mighty men' (IV, 2, 7-20).

[2] Another example occurs in I, 4, when Firke and Hodge both 'offer to go' unless their master will find employment for Hans.

masters and apprentices under the heading of Cruelty. Several companies in the City, he says, are busy cultivating the arts of cruelty which they apply both during the seven years apprenticeship and afterwards. 'When they have fared hardly with you by indenture and, like your beasts which carry you, have patiently borne all labour and all wrongs you could lay upon them', they then 'lay their heads together in conspiracy' to prevent their men becoming masters. So it is 'as if trades, that were ordained to be communities, had lost their first privileges, and were now turned to monopolies'. Perhaps Dekker had created Eyre as a model master and thus as a rebuke to others, but there is no sense in the play of any similar social criticism making itself felt by either explicit or implicit means. It would seem that the evils he denounced seven years later had impressed themselves upon him in the meantime. And it would be surprising if time had not worked in this way, for Dekker's does not seem to have been the most fortunate of lives.

True, it was quite a long one. Only approximate dates are known: he was probably born in 1570 and died in 1637. On very uncertain grounds, he is said to have gone to Merchant Taylors' School, though his name is not in the school register. It may be that he educated himself more effectively than any school managed to do, and that he is speaking on his own behalf in the poem *Dekker His Dream* when he makes a character refer to

> My private readings, which more schooled my soul
> Than tutors, when they sternliest did control
> With frowns or rods.

He is fond in his plays of introducing foreigners who speak a kind of Dutch, French or Welsh; but this is not enough to assure us either that he travelled a great deal or that he had an advanced education. The likelihood is that he picked up a working knowledge of these languages and was a good observer. Certainly he worked hard. Twenty-four plays and thirteen prose works have survived and there are contemporary references to others. He received a good salary for the earlier plays—we know this from the diary of Phillip Henslowe, the theatre manager. Nevertheless, he spent seven years (1613–20) in prison, probably for debt, and *Dekker His Dream* begins with a poignant reference to the unhappiness of this time:

the bed on which seven years I lay dreaming was filled with thorns instead of feathers; my pillow a rugged flint; my chamberfellows (sorrows that day and night kept me company) the very or worse than the very infernal furies.

In many of his more personal utterances he shows himself fully aware of the frustrations of a writer's life and the bitterness they can produce. 'I am mad to see men scholars in the broker's shop, and dunces in the mercer's', says a character in *Old Fortunatus* (1, 2, 133). 'The labours of writers are as unhappy as the children of a beautiful woman, being spoilt by ill nurses, within a month or two after they come into the world.' So he complained in the Preface to *The Whore of Babylon*, the initial failure of which he blamed on the bad performers. And more bitterly still, he writes: 'To come to the press is more dangerous than to be pressed to death, for the pains of those tortures last but a few minutes, but he that lives upon the rack in print hath his flesh torn off by the teeth of envy and calumny, even when he means nobody any hurt in his grave.... Take heed of critics: they bite like fish at anything, especially at books.' Dekker's bitterness or depression never flattens his prose, but the author of *The Shoemaker's Holiday* was, one would say, such a merry fellow that it comes as a surprise to find these complaints in his writing at all.

He is, of course, a far more *complete* writer for having this darker side to him. *The Shoemaker's Holiday* is still his best-known work, but I do not think that on the knowledge of that play alone, many readers would take Dekker all that seriously as a writer or a man. A light-hearted entertainer, a good observer, a dramatist with a lively command of language, we might say, but hardly a great deal more. As we have seen, his range was greater than that, and there is more to him still. The kindly humour of *The Shoemaker's Holiday*, for example, does not lead one to expect the satirical sharpness of his play *Satiromastix*, with its well-sustained mockery of Ben Jonson; vigorous rather than neat, but still pointed and unsparing enough. Nor would one credit him with the ability to turn the screw of his attack on moneyed meanness with the sort of irony we see here:

Let the times be dear though the grounds be fruitful, and the markets kept empty though your barns (like cormorants' bellies) break their button-holes; and rather than any of Poverty's soldiers, who now range up and down the kingdom, should have bread to relieve them: I charge you upon your allegiance to hoard up your corn till it be musty, and then bring it forth to infect these needy Barbarians, that the rot, scurvy or some other infectious disease may run through the most part of their enfeebled army. Or... let mice and rats rather be feasted by you and fare well in your garners, than the least and weakest amongst Poverty's starved infantry should get but one mouthful. Let them leap at crusts, it shall be sport enough for us and our wealthy subjects about us, to laugh at them whilst they nibble at the bait and yet be choked with the hook.

(*Work for Armourers*, 1609)

This is from a proclamation by the Queen of Gold and Silver, who is leading the rich men's war against the poor. The technique (saying the opposite of what the author really means and feels) plays for ridicule, as well as for the moral condemnation of the decent man.

Introduction

Perhaps if he had worked a play instead of a pamphlet out
of the depth of feeling that went into *Work for Armourers,*
Dekker might have written a successful tragedy (and it would
have been unique in kind, for the suffering of poor people
would have been its subject). As it is, he is essentially a writer
of comedies, though sometimes these touch seriousness and
just occasionally maintain it with some strength. The best
play to sample for this is Part II of *The Honest Whore* (included
in the Mermaid edition of Dekker). Here, particularly in the
third act, we find a concern with real problems. The comic
scenes illustrate the traditional doctrine that a wife must be
subject to her husband; but then in a serious scene the other
side is presented. The convention which allows a man the
sexual licence for which it condemns a woman is attacked
with subtlety and feeling. There is a moving passage too in
which the sufferings of a woman married to a wastrel are
tensely and eloquently dramatised. The play does not rise to
tragedy, but it is good enough sometimes to call to mind the
Shakespeare of *Measure for Measure.*

In other plays it is the Shakespeare of *The Winter's Tale*
that we think of. *Patient Grissil,* in many ways absurd and
monstrous in that we are supposed to see the trials imposed
on a good woman as somehow justified by the god-like
status of the King who devises them, can nevertheless be
moving; especially towards the end, where a rare serenity is
attained. There is a grace and a graciousness over it all; a
sort of dream-like quality which recalls the endings of
Shakespeare's last plays: Hermione being reconciled to
Leontes in *The Winter's Tale,* or Prospero blessing Miranda
and Ferdinand after their trials in *The Tempest.* Sometimes
the language rises to it:

GRISSIL Blessing distil on you like morning dew;
My soul knit to your souls, knows you are mine.

MARQUESS They are, and I am thine: Lords, look not strange.
These two are they, at whose birth envy's tongue
Darted envenomed stings; these are the fruit
Of this most virtuous tree...
My Grissil lives, and in the book of fame
All worlds in gold shall register her name. (v, 2, 196–201; 207–8)

V

But it is by *The Shoemaker's Holiday* that Dekker has been
generally remembered. There are good reasons for its long
survival: the vivid background of London life, the fresh
tenderness and humour of some of the love scenes, and,
lighting up the whole play, the kindly energy of Simon Eyre,
protector of lovers, champion of shoemakers, the middle-
aged madcap whom age has not withered or high office
staled. He must be one of the most likeable characters in
literature. From his first appearance to his last we see him
coming forward on behalf of others. At the beginning it is
to plead for Rafe; at the end it is to ask a royal favour for his
fellow-craftsmen. When the plea for Rafe's exemption from
the army is refused, he has no resentment, and can use his
cheerfulness and wit to keep up the spirits of everyone else.
There never was a man with less of a chip on his shoulder;
this too is very un-modern, for our present-day literature
abounds in resentfulness. Not that Eyre is given to any soft-
centred compliancy: his defence of Rose and Lacy illustrates
that. He also has an eye to a sharp bit of business, for the
bargain which makes his fortune is procured on borrowed
money and (after all) false credit. Eyre dresses up in alderman's gown and other finery so as to impress the Skipper
and have his 'earnest' accepted.[1] Nor is he such a starry-eyed

---

[1] Dekker has played this down: in his source-book the trick is much more
clearly exposed (see below, p. 138).

employer, however free and convivial with his men. He jollies them along most of the time, but does not let them forget what they owe to him:

Have I not ta'en you from your selling tripes in Eastcheap and set you in my shop, and made you hail-fellow with Simon Eyre the shoe-maker? (II, 3, 67–9)

—all said, of course, in the usual tone of jovial banter. A few lines farther on we have another example of his genial control of the situation. He has made peace with his rebellious journeymen and now orders a dozen cans of beer to celebrate:

FIRKE A dozen cans? Oh brave! Hodge, now I'll stay.
EYRE *(aside to the boy)* An the knave fills any more than two, he pays for them. (77–8)

It is this mixture of generosity and realism that makes a naturally endearing character also a strong one. With liberality goes thrift, with the impulsive and boisterous high spirits go sharpness and authority. His way with his wife is a similar compound of banter and affection. He calls her all sorts of names—'powder-beef quean', 'queen of clubs', 'rubbish' and 'kitchen-stuff'—yet there is no doubt she is his other half, for in all his triumph and excitement he is constantly fussing with his 'Lady Madgy' to make sure that she is sharing it. This is the spirit that fills the play. Eyre is on stage in less than half the scenes, but his goodness and gaiety provide the keynote throughout.

Firke, Hodge and Margery all sing more or less to his tune: they have a similar fund of energy and goodheartedness. Mistress Eyre is at her best in III, 2. We have her, buxom and red-faced, impatient to know what is happening at the Guild-hall, and exercising her 'posh' voice and consequential manner when she finds herself addressed as 'Mistress Shrieve'. There is also her touching dialogue with the war-wounded Rafe, her sense of humour and pity being equally stirred, yet with half her mind still on her own affairs and her husband's. This is her best scene. Firke's is probably iv, 4, where he plays a fine old game with the fuming Lord Mayor and Earl of Lincoln. But he is a useful comic throughout, with a style of his own and a liveliness as inexhaustible as his master's.

The play abounds in good secondary acting parts. There is Sybil: too jolly a girl for all tastes but a good foil to the more maidenly Rose; Rafe, whose dazed interview with Hammon's servant is so well caught (iv, 2); Hammon himself, who having begun life in the play as an opportunist, a playboy and a nuisance, eventually gains sympathy as luck goes persistently against him, retiring with dignity and in sorrow so that for a moment he nearly overtips the balance and makes us less jubilant about the tough line which the shoemakers have so successfully taken against him. Jane, his intended bride, is also a part for acting, and not merely a fill-in. Her scene with Hammon (iii, 4) is probably the most affecting in the play, ranging from a rather charming playfulness, through per-plexity and distress, to genuine grief. Unhappiness never stays on the stage for long, but while it is there no cheap or false touch disfigures the treatment of it.

Lincoln and Oatley, Lacy and Rose, are perhaps less fully flavoured characters, and their part of the play, the love-plot, certainly has less life than the rest. Even so, the lovers are a pleasant couple and not as insipid as the romantic leads in most modern comedies. Rose is a strong-willed girl who can hold parents and suitors at bay when she wants, and whose wit moves more quickly than her lover's at a critical moment later on (iv, 3). Lacy is no conventional hero, for his desertion of class and colours would brand him as a cad in a more conventional play. The King excuses him, quite seeing that love is a very understandable reason for leaving the army to get on as best it can; and Lacy's unsqueamish and spirited

Introduction

adoption of a working-class existence is presented as a good, manly characteristic compared with the niggling class-consciousness of his uncle and father-in-law.

'Class' is probably the nearest thing to a theme in *The Shoemaker's Holiday*. It is a habit, often a fruitful one, of modern criticism to look beyond the story and characters of a play to find some underlying concern which gives it unity and depth. Sometimes distortion goes on in the process, and it would not be very fruitful to hunt about for 'significant passages' in the sort of entertainment Dekker has written here. But consciousness of class or indifference to it are amongst the attitudes that he most repeatedly dramatises. Lincoln's insistence that Rose is 'too too base' in birth represents no doubt the normal sentiments of someone who has such a name as Lacy to save from contamination. But the Lord Mayor's inverted snobbery is just as marked. He stands for the barricading of his own class as proudly as any aristocrat: he wants a son-in-law who follows a trade, a gentleman-citizen and no 'silken fellow' with a fancy pedigree. Eyre himself has some of this class feeling: 'marry me with a gentleman grocer like my Lord Mayor your father' is his advice to Rose. Mistress Eyre too, when she goes up in the world, becomes very conscious of her station: in dress, speech and manners she will try to 'keep up'. But that is there to be laughed at, and so is most of this awareness of class. Lacy is indifferent to it, and the King speaks out unequivocally against any social exclusiveness where love is concerned:

> Dost thou not know that love respects no blood,
> Cares not for difference of birth or state?   (v, 5, 108-9)

he says to the Earl of Lincoln. Going with this is the essential spirit of the play, embodied in Eyre, who is his true self whether with his King or his apprentices. 'Prince am I none, yet am I princely born' is his cry: this was an old saying of

18

the shoemakers, but Eyre's proud reiteration of it springs from that sense of personal dignity and worth that cuts across classes and makes snobbery barren and absurd.

Still, *The Shoemaker's Holiday* is not a 'problem play' or a social drama with a thesis. It is an entertainment in verse and prose, and criticism of its quality must base itself on that. On recognition, too, of the stagecraft involved, for Dekker is always workmanlike. He develops three plots: the love story of Lacy and Rose, the history of the advancement of Simon Eyre, and the separation and reuniting of Jane and Rafe. All of them are set moving in the first scene, and with a sure touch each is directed towards its climax. The climax of one plot is neatly made to involve that of another (when Rafe finds his Jane, Lacy has won his Rose and the two are about to 'chop up the matter at the Savoy', while their frustrated elders find themselves hindering the wrong couple, only to be defeated afresh at their second attempt when the good fortune of the lovers coincides with the final triumph of Simon Eyre). In the meantime the dramatist has been working resourcefully to keep the entertainment alive and varied. Knockabout is balanced by wit and a kind of verbal brilliance that are delightful in themselves; the earthy zest of Firke and the rest is offset by the more delicate language of Rose. Nor is all the world tediously playing holiday, for the sight of Rafe with his crutch and single leg cannot have been found amusing in any age, nor can the distress of Jane when she sees her Rafe's name on the list of those killed in the wars. Her unhappiness and the wretched pesterings of Hammon are dramatised with imagination and sympathy; and in actual stage production the episodes are doubly effective.

So they turned out to be in a production of the play in 1964 at the Mermaid Theatre, London. In this, the scene which came most freshly to life was III, 2, where Rafe returns 'being lame' from France. The affection and embarrassment of Mistress Eyre and the others, the joviality and the pity, were touchingly brought out; it would take a very imaginative reading of the printed page to make one realise just how supple and lifelike this scene can be. On the other hand, the production as a whole did not seem to me to be at all satisfactory. There was plenty of laughter in the theatre, but the vitality and wit of language accounted for little of it. Indeed, words were given little time to register or even to be fully audible. Instead the poor old play was revitalised by an injection of what the *Guardian* critic called 'proper ho-ho' (meaning back-slapping, ear-splitting heartiness); and worse. Chamber pots are always good for a laugh, and when emptied over the stage from a balcony can (apparently) be considered very amusing indeed. Firke, more or less in the wings, was supposed to urinate noisily; a bird was supposed to drop something which made Rose shriek with dismay at the end of her first scene with Sybil; and the King was presented as a coquettish old queen who could not decide which crown to wear. Such 'elegant elaborations of the text', said the *Guardian* critic, 'are not what I would call funny'. But he seemed to be alone, for most critics were delighted and the *Daily Telegraph* carried the headline 'Dekker lives again'.

Worse still, Eyre was not allowed the dignity and serious depth of character that are his, and that he must have if the play is not to be coarsened. But it is perhaps not without significance that the play should be offered in this coarsened form by a modern company that, after all, specialises in plays of the period; and accepted by an audience that must, after all, be more 'cultivated' than most. This happens in an age which feels itself to be sophisticated, and all the more so for being able to enjoy a 'simple, Elizabethan romp'.

We began by speaking of the kind of sophistication the

play does *not* possess (while we ourselves do); but it now becomes apparent that another kind of sophistication, and a rather better kind (that we do not so evidently have), is precisely what does have to be attributed to it. For the dramatist's skill in blending these several elements and working them together side by side is considerable, and so is the flexibility and responsiveness required of his audience. The story itself (particularly the love story) is very much the conventional assembly of disguises and confusions. Such conventions were readily acceptable to the Elizabethan audiences presumably because, like a pantomime-plot, it was part of the anticipated fun of play-going, and because it was in any case essentially a vehicle, something which allowed spectators to concentrate on the humour—which is mostly verbal. This was in 1599, and the audience had got past the stage of gaping at mere event and being impressed by rant. They had, in other words, achieved very genuine sophistication.

Still, whatever admiration one has for them and their author, it has to be admitted that the quality of writing in *The Shoemaker's Holiday* is not always distinguished. The most noticeable characteristic is probably how the words come to life when blank verse ends and prose takes over. This is generally so with Dekker. As a writer of prose he had few superiors in what was, after all, a great age. His sentences can skip about with an easy colloquial freedom, or they can move with biblical solemnity and splendour. Always he has an ear for the rhythm of a sentence:

Let us awhile leave kingdoms and enter into cities. Sodom and Gomorrah were burnt to the ground with brimstone that dropped its flakes from heaven: a hot and dreadful vengeance. Jerusalem hath not a stone left upon another of her first glorious foundation: a heavy and fearful downfall. Jerusalem, that was God's own dwelling house, the school where those Hebrew lectures which he himself read were taught, the very nursery where the Prince of heaven was brought up: that Jerusalem whose rulers were princes and whose citizens were like the sons of kings, whose temples were paved with gold and whose houses stood like rows of tall cedars: that Jerusalem is now a desert. It is unhallowed and untrodden; no monument is left to show it was a city, but only the memorial of the Jews' hard-heartedness in making away their saviour. It is now a place for barbarous Turks and poor despised Grecians; it is rather now (for the abominations committed in it) no place at all.

(The Seven Deadly Sins of London)

His blank verse is generally workmanlike, but he is not a great poet by any means. After a Preface 'To the Reader', written, like the quotation from *The Seven Deadly Sins*, with marvellous grace and dignity, he opens his poem *Dekker His Dream* with a short rhymed introduction:

> When down the Sun his golden beams had laid,
> And at the western inn his journey stayed,
> Thus sleep the eyes of man and beast did seize,
> Whilst he gave light to the Antipodes.
> I slept with others, but my senses streamed
> In frightful forms, for a strange dream I dreamed.

One wonders what has happened to his ear. Nothing as crude as this occurs in *The Shoemaker's Holiday*, but the verse is rarely anything more than a serviceable, staple commodity; and it is much more prosaic than the prose. An example would be I, 3 and 4, where Lacy has a blank-verse speech of twenty-four lines followed by a speech of Eyre's in prose. The verse is graceful enough and perfectly clear, but the movement of the lines is uninteresting and the language lacks colour. The prose by comparison is rhythmical and imaginative:

Where be these boys, these girls, these drabs, these scoundrels? They wallow in the fat brewis of my bounty and lick up the crumbs of my table, yet will not rise to see my walks cleansed. Come out, you powder-beef queans!   (I, 4, 1–5)

But here, of course, Dekker is in his element. It is not simply that he is better in prose than in verse, but that the verse generally goes to the ladies and gentlemen, and the prose to the working-class entertainers. Eyre's prose is charged with the vitality of his generous character and the energy of colloquial speech which Dekker could draw on through him.

These exactly matching energies of language and good-heartedness make Dekker's comedy the worth-while thing it is. Ultimately, the lightest of entertainments has its influence upon a society for good or bad. There is no matter in *The Shoemaker's Holiday* that is not wholesome, no scene where inner deadness and manufactured cleverness are suspect. There is little subtlety, little irony, or satire: and sometimes we feel this as a limitation to our enjoyment. 'Psha! there's no possibility of being witty without a little ill-nature: the malice of a good thing is the barb that makes it stick.' One part of us says 'Amen' to that. But the speaker is Lady Sneerwell (*School for Scandal*, I, I) and her character is what her name proclaims it to be. At any rate, *The Shoemaker's Holiday* manages without ill-nature or malice, and if it does not offer Lady Sneerwell's kind of wit, then there are high spirits in plenty, zest and a sense of decent communal fellowship to compensate. Deeper problems and more complicated emotions intervene at critical points in life, but in the meantime these virtues take us a long way. In *His Dream* Dekker says that life is a voyage to the happiness of heaven: 'Books are pilots in such voyages: would mine were but one point of the compass, for any man to steer well by.' It is doubtful whether that particular book, with its vision of judgment and hell-fire as the ultimate deterrent, will help to pilot any man born in our times. But the very simple goodness and energy of Eyre and his shoemakers might.

A NOTE ON THE TEXT

The text of this edition is based on that given by Professor Fredson Bowers (*The Dramatic Works of Thomas Dekker*, vol. 1, Cambridge University Press, 1953). Spelling and punctuation have been modernised by the present editor.

PERSONS

KING OF ENGLAND
EARL OF LINCOLN
EARL OF CORNWALL
SIR ROGER OATLEY, Lord Mayor of London
SIMON EYRE, shoemaker and afterwards Lord Mayor
ROWLAND LACY, nephew to Lincoln, afterwards disguised
  as Hans Meulter
ASKEW, cousin to Lacy
HAMMON, a city gentleman
WARNER, a cousin to Hammon
MASTER SCOTT, a friend to Oatley
HODGE (also called Roger), foreman to Eyre
FIRKE, journeyman to Eyre
RAFE DAMPORT, journeyman to Eyre
LOVELL, servant to the King
DODGER, parasite to Lincoln
Dutch Skipper
Boy, apprentice to Eyre
Boy, servant to Oatley

MARGERY, wife to Eyre
ROSE, daughter to Oatley
JANE, wife to Rafe Damport
SYBIL, maid to Rose

Noblemen, soldiers, huntsmen, shoemakers, apprentices,
  servants

To all good Fellows, Professors of the Gentle Craft; of what degree soever

Kind gentlemen and honest boon companions, I present you here with a merry conceited comedy called *The Shoemaker's Holiday*, acted by my Lord Admiral's Players this present Christmas before the Queen's most excellent Majesty. For the mirth and pleasant matter, by Her Majesty graciously accepted, being indeed no way offensive. The Argument of the play I will set down in this Epistle: Sir Hugh Lacy, Earl of Lincoln, had a young gentleman of his own name, his near kinsman, that loved the Lord Mayor's daughter of London; to prevent and cross which love the Earl caused his kinsman to be sent Colonel of a company into France: who resigned his place to another gentleman, his friend, and came disguised like a Dutch shoemaker to the house of Simon Eyre in Tower Street, who served the Mayor and his household with shoes. The merriments that passed in Eyre's house, his coming to be Mayor of London, Lacy's getting his love, and other accidents; with two merry Three-mans songs. Take all in good worth that is well intended, for nothing is purposed but mirth. Mirth lengtheneth long life; which, with other blessings, I heartily wish you. Farewell.

The first Three-mans Song

O the month of May, the merry month of May,
So frolic, so gay, and so green, so green, so green:
Oh and then did I unto my true love say,
'Sweet Peg, thou shalt be my summer's queen'.

Now the nightingale, the pretty nightingale,
The sweetest singer in all the forest's choir,
Entreats thee, sweet Peggy, to hear thy true love's tale:
Lo, yonder she sitteth, her breast against a briar.

But Oh I spy the cuckoo, the cuckoo, the cuckoo:
See where she sitteth; come away, my joy.
Come away, I prithee; I do not like the cuckoo
Should sing where my Peggy and I kiss and toy.

Oh the month of May, the merry month of May,
So frolic, so gay, and so green, so green, so green:
And then did I unto my true love say,
'Sweet Peg, thou shalt be my summer's queen'.

The second Three-mans Song

This is to be sung at the latter end

Cold's the wind and wet's the rain,
    Saint Hugh be our good speed:
Ill is the weather that bringeth no gain,
    Nor helps good hearts in need.

Trowl the bowl, the jolly nut-brown bowl,
    And here, kind mate, to thee:
Let's sing a dirge for Saint Hugh's soul,
    And down it merrily.

Down a-down, hey down a-down,
   Hey derry derry down a-down, *Close with the tenor boy*
Ho, well done, to me let come,
   Ring compass gentle joy.

Trowl the bowl, the nut-brown bowl,
   And here kind &c. *as often as there be men to drink*

       At last when all have drunk, this verse:

Cold's the wind and wet's the rain,
   Saint Hugh be our good speed:
Ill is the weather that bringeth no gain,
   Nor helps good hearts in need.

The Prologue, as it was pronounced before the Queen's Majesty

As wretches in a storm, expecting day,
With trembling hands and eyes cast up to heaven,
Make prayers the anchor of their conquered hopes:
So we, dear goddess, wonder of all eyes,
Your meanest vassals, through mistrust and fear
To sink into the bottom of disgrace
By our imperfect pastimes, prostrate thus
On bended knees, our sails of hope do strike,
Dreading the bitter storms of your dislike.
Since then, unhappy men, our hap is such,
That to ourselves no help can bring,
But needs must perish if your saint-like ears,
Locking the temple where all mercy sits,
Refuse the tribute of our begging tongues.
Oh grant, bright mirror of true chastity,
From those life-breathing stars, your sun-like eyes,
One gracious smile: for your celestial breath
Must send us life, or sentence us to death.

29

A pleasant Comedy of the Gentle Craft

ACT I

SCENE I

A street in London. Enter the LORD MAYOR
(SIR ROGER OATLEY) *and the* EARL OF LINCOLN.

LINC. My Lord Mayor, you have sundry times
    Feasted myself and many courtiers more:
    Seldom or never can we be so kind
    To make requital of your courtesy.
    But leaving this, I hear my cousin Lacy   5
    Is much affected to your daughter Rose.
L.MA. True, my good lord, and she loves him so well
    That I mislike her boldness in the chase.
LINC. Why, my Lord Mayor? Think you it then
      a shame
    To join a Lacy with an Oatley's name?   10
L.MA. Too mean is my poor girl for his high birth.
    Poor citizens must not with courtiers wed,
    Who will in silks and gay apparel spend
    More in one year than I am worth by far.
    Therefore your honour need not doubt my girl.   15
LINC. Take heed, my Lord; advise you what you do.
    A verier unthrift lives not in the world
    Than is my cousin. For I'll tell you what:
    'Tis now almost a year since he requested
    To travel countries for experience.   20
    I furnished him with coin, bills of exchange,
    Letters of credit, men to wait on him,
    Solicited my friends in Italy

Well to respect him. But to see the end:
Scant had he journeyed through half Germany   25
But all his coin was spent, his men cast off,
His bills embezzled; and my jolly cos,
Ashamed to show his bankrupt presence here,
Became a shoemaker in Wittenberg,
A goodly science for a gentleman   30
Of such descent! Now judge the rest by this:
Suppose your daughter have a thousand pounds,
He did consume me more in one half year.
And make him heir to all the wealth you have,
One twelve month's rioting will waste it all.   35
Then seek, my Lord, some honest citizen
To wed your daughter to.

L.MA.   I thank your Lordship.
 Well, fox, I understand your subtlety.   *Aside.*
 As for your nephew, let your Lordship's eye
 But watch his actions; and you need not fear,   40
 For I have sent my daughter far enough.
 And yet your cousin Rowland might do well
 Now he hath learned an occupation.
 (And yet I scorn to call him son-in-law).   *Aside.*
LINC. Ay, but I have a better trade for him.   45
 I thank his Grace, he hath appointed him
 Chief Colonel of all those companies
 Mustered in London and the shires about
 To serve his Highness in those wars in France.
 See where he comes—

Enter LOVELL, LACY *and* ASKEW

     Lovell, what news with you?   50
LOVELL My Lord of Lincoln, 'tis his Highness' will
 That presently your cousin ship for France

    With all his powers. He would not for a million
    But they should land at Dieppe within four days.
LINC. Go certify his Grace it shall be done.   *Exit Lovell.* 55
    Now, cousin Lacy. In what forwardness
    Are all your companies?
LACY   All well prepared.
    The men of Hertfordshire lie at Mile End,
    Suffolk and Essex train in Tothill Fields,
    The Londoners and those of Middlesex,   60
    All gallantly prepar'd in Finsbury,
    With frolic spirits long for the parting hour.
L.MA. They have their imprest, coats and furniture,
    And if it please your cousin Lacy come
    To the Guildhall, he shall receive his pay;   65
    And twenty pounds besides my brethren
    Will freely give him, to approve our loves
    We bear unto my Lord your uncle here.
LACY I thank your honour.
LINC.   Thanks, my good Lord Mayor.
L.MA. At the Guildhall we will expect your coming. *Exit.* 70
LINC. To approve your loves to me? No! Subtlety!
    Nephew, that twenty pound he doth bestow
    For joy to rid you from his daughter Rose.
    For, cousins both, now here are none but friends,
    I would not have you cast an amorous eye   75
    Upon so mean a project as the love
    Of a gay, wanton, painted citizen.
    I know this churl even in the height of scorn
    Doth hate the mixture of his blood with thine.
    I pray thee do *thou* so. Remember, cos,   80
    What honourable fortunes wait on thee.
    Increase the King's love which so brightly shines
    And gilds thy hopes. I have no heir but thee:

And yet not thee if, with a wayward spirit,
Thou start from the true bias of my love.   85
LACY   My Lord, I will—for honour, not desire
  Of land or livings, or to be your heir—
  So guide my actions in pursuit of France
  As shall add glory to the Lacys' name.
LINC. Cos, for those words here's thirty Portuguese.   90
  And nephew Askew, there's a few for you.
  Fair Honour in her loftiest eminence
  Stays in France for you till you fetch her thence.
  Then, nephews, clap swift wings on your designs.
  Be gone, be gone: make haste to the Guildhall.   95
  There presently I'll meet you. Do not stay:
  Where Honour beckons, Shame attends delay.   *Exit.*
ASKEW How gladly would your uncle have you gone!
LACY   True, cos, but I'll o'erreach his policies.
  I have some serious business for three days,   100
  Which nothing but my presence can dispatch.
  You therefore, cousin, with the companies,
  Shall haste to Dover. There I'll meet with you,
  Or, if I stay past my prefixèd time,
  Away for France: we'll meet in Normandy.   105
  The twenty pounds my Lord Mayor gives to me
  You shall receive, and these ten Portuguese,
  Part of mine uncle's thirty. Gentle cos,
  Have care to our great charge. I know your wisdom
  Hath tried itself in higher consequence.   110
ASKEW Cos, all myself am yours. Yet have this care:
  To lodge in London with all secrecy.
  Our uncle Lincoln hath (besides his own)
  Many a jealous eye, that in your face
  Stares only to watch means for your disgrace.   115
LACY   Stay cousin—who be these?

Act I, Scene 1

Enter SIMON EYRE, *his* WIFE MARGERY, HODGE,
FIRKE, JANE *and* RAFE *with a piece.*

EYRE Leave whining, leave whining! Away with this
whimpering, this puling, these blubbering tears and these
wet eyes! I'll get thy husband discharged, I warrant thee,
sweet Jane; go to.   120

HODGE Master, here be the captains.

EYRE Peace, Hodge. Hushed, ye knaves, hushed!

FIRKE Here be the cavaliers and the colonels, master.

EYRE Peace, Firke. Peace, my fine Firke. Stand by with
your pishery-pashery, away! I am a man of the best 125
presence. I'll speak to them and they were Popes. (*to Lacy
and Askew*) Gentlemen, captains, colonels, commanders!
Brave men, brave leaders, may it please you to give me
audience. I am Simon Eyre, the mad shoemaker of Tower
Street. This wench with the mealy mouth that will never 130
tire is my wife, I can tell you. Here's Hodge, my man
and my foreman. Here's Firke, my fine firking journeyman. And this is blubbered Jane. All we come to be
suitors for this honest Rafe. Keep him at home and,
as I am a true shoemaker, and a gentleman of the Gentle 135
Craft, buy spurs your self and I'll find ye boots these seven
years.

WIFE Seven years, husband?

EYRE Peace, midriff, peace. I know what I do. Peace.

FIRKE Truly, master Cormorant, you shall do God good 140
service to let Rafe and his wife stay together. She's a young
new-married woman. If you take her husband away from
here anight, you undo her. She may beg in the day-time,
for he's as good a workman at a prick and an awl as any is
in our trade.   145

JANE Oh let him stay, else I shall be undone.

34

FIRKE   Ay, truly, she shall be laid at one side like a pair of
old shoes else, and be occupied for no use.

LACY   Truly my friends, it lies not in my power.
The Londoners are pressed, paid and set forth   150
By the Lord Mayor. I cannot change a man.

HODGE   Why then, you were as good be a Corporal as a
Colonel if you cannot discharge one good fellow. And I
tell you true: I think you do more than you can answer, to
press a man within a year and a day of his marriage.   155

EYRE   Well said, melancholy Hodge. Gramercy, my fine
foreman.

WIFE   Truly, gentlemen, it were ill done for such as you to
stand so stiffly against a poor young wife, considering her
case. She is but new married (but let that pass). I pray,   160
deal not roughly with her. Her husband is a young man
and but newly entered—but let that pass.

EYRE   Away with your pishery-pashery, your pols and your
edipols! Peace, midriff! Silence, Cicely Bumtrinket: let
your head speak!   165

FIRKE   Yea, and the horns too, master.

EYRE   Too soon, my fine Firke, too soon. Peace, scoundrels.
See you this man, captains? You will not release him—well,
let him go. He's a proper shot; let him vanish. Peace,
Jane, dry up thy tears; they'll make his powder dankish.   170
Take him, brave men! Hector of Troy was an hackney
to him, Hercules and Termagant scoundrels. Prince
Arthur's Round Table, by the Lord of Ludgate, ne'er fed
such a tall, such a dapper swordsman. By the life of
Pharaoh, a brave, resolute swordsman! Peace, Jane. I say   175
no more, mad knaves.

FIRKE   See, see Hodge, how my master raves in commenda-
tion of Rafe.

HODGE   Rafe, thou art a gull, by this hand, an thou goest not.

ASKEW   I am glad, good Master Eyre, it is my hap   180
   To meet so resolute a soldier.
   Trust me, for your report and love to him
   A common slight regard shall not respect him.
LACY   Is thy name Rafe?
RAFE   Yes, sir.
LACY   Give me thy hand.
   Thou shalt not want, as I am a gentleman.   185
   Woman, be patient. God, no doubt, will send
   Thy husband safe again. But he must go:
   His country's quarrel says it shall be so.
HODGE   Thou art a gull, by my stirrup, if thou dost not go,
   I will not have thee strike thy gimlet into these weak 190
   vessels. Prick thine enemies, Rafe.

Enter DODGER.

DODGER   My Lord, your uncle on the Tower Hill
   Stays with the Lord Mayor and the Aldermen,
   And doth request you with all speed you may
   To hasten thither.   195
ASKEW   Cousin, let us go.
LACY   Dodger, run you before. Tell them we come.
   This Dodger is mine uncle's parasite,   *Exit Dodger.*
   The arrant'st varlet that e'er breathed on earth.
   He sets more discord in a noble house   200
   By one day's broaching of his pickthank tales
   Than can be salved again in twenty years.
   And he, I fear, shall go with us to France
   To pry into our actions.
ASKEW   Therefore, cos,
   It shall behove you to be circumspect.   205
LACY   Fear not, good cousin. Rafe, hie to your colours.
                       Exeunt Lacy and Askew.

RAFE   I must, because there is no remedy.
   But gentle master and my loving dame,
   As you have always been a friend to me,
   So in mine absence think upon my wife.   210

JANE   Alas, my Rafe.

WIFE   She cannot speak for weeping.

EYRE   Peace, you cracked groats, you mustard tokens!
   Disquiet not the brave soldier. Go thy ways, Rafe.

JANE   Ay ay, you bid him go. What shall I do when he is 215
   gone?

FIRKE   Why, be doing with me, or my fellow Hodge Be
   not idle.

EYRE   Let me see thy hand, Jane. This fine hand, this white
   hand, these pretty fingers must spin, must card, must work. 220
   Work, you bombast cotton-candle-quean, work for your
   living, with a pox to you. Hold thee, Rafe: here's five
   sixpences for thee. Fight for the honour of the Gentle
   Craft, for the gentlemen shoemakers, the courageous
   cordwainers, the flower of Saint Martin's, the mad knaves 225
   of Bedlam, Fleet Street, Tower Street and Whitechapel.
   Crack me the crowns of the French knaves, a pox on them!
   Crack them! Fight, by the Lord of Ludgate. Fight, my
   boy.

FIRKE   Here, Rafe, here's three two-pences. Two carry into 230
   France; the third shall wash our souls at parting (for sorrow
   is dry). For my sake, firk the *Basa mon cues*.

HODGE   Rafe, I am heavy at parting, but here's a shilling for
   thee. God send thee to cram thy slops with French crowns,
   and thy enemies' bellies with bullets.   235

RAFE   I thank you, master, and I thank you all.
   Now, gentle wife, my loving, lovely Jane,
   Rich men at parting give their wives rich gifts,
   Jewels and rings to grace their lily hands.

Thou know'st our trade makes rings for women's heels: 240
Here, take this pair of shoes cut out by Hodge,
Stitched by my fellow Firke, seamed by myself,
Made up and pinked with letters for thy name.
Wear them, my dear Jane, for thy husband's sake,
And every morning when thou pull'st them on   245
Remember me and pray for my return.
Make much of them; for I have made them so,
That I can know them from a thousand mo.

Sound drum. Enter LORD MAYOR, LINCOLN, LACY, ASKEW,
DODGER *and soldiers. They pass over the stage.* RAFE *falls in
amongst them.* FIRKE *and the rest cry Farewell, etc., and so exeunt.*

SCENE 2

*A garden at Old Ford, Sir Roger Oatley's house outside London.
Enter* ROSE *alone, making a garland.*

ROSE Here sit thou down upon this flowery bank,
And make a garland for thy Lacy's head.
These pinks, these roses and these violets,
These blushing gillyflowers, these marigolds,
The fair embroidery of his coronet,   5
Carry not half such beauty in their cheeks
As the sweet countenance of my Lacy doth.
Oh my most unkind father! Oh my stars,
Why loured you so at my nativity
To make me love, yet live robbed of my love?   10
Here as a thief am I imprisonèd
For my dear Lacy's sake, within those walls
Which by my father's cost were builded up
For better purposes. Here must I languish
For him that doth as much lament, I know,   15

Enter SYBIL.

Mine absence, as for him I pine in woe.

SYBIL   Good morrow, young mistress. I am sure you make
that garland for me, against I shall be Lady of the Harvest.

ROSE   Sybil, what news at London?

SYBIL   None but good. My Lord Mayor your father, and 20
Master Philpot your uncle, and Master Scott your cousin,
and Mistress Frigbottom by Doctors' Commons, do all, by
my troth, send you most hearty commendations.

ROSE   Did Lacy send kind greetings to his love?

SYBIL   Oh yes, out of cry. By my troth, I scant knew him. 25
Here 'a wore a scarf and here a scarf, here a bunch of
feathers, and here precious stones and jewels and a pair of
garters. Oh monstrous! Like one of our yellow silk
curtains at home here in Old Ford House, here in Master
Bellymount's chamber. I stood at our door in Cornhill, 30
looked at him, he at me indeed, spake to him, but he not
to me, not a word. 'Marry, go up', thought I, 'with a
wanion.' He passed by me as proud. 'Marry, foh! Are
you grown humorous?' thought I, and so shut the door,
and in I came.   35

ROSE   Oh Sybil, how dost thou my Lacy wrong!
My Rowland is as gentle as a lamb.
No dove was ever half so mild as he.

SYBIL   Mild? Yea, as a bushel of stamped crabs. He looked
upon me as sour as verjuice. 'Go thy ways!' thought I. 40
'Thou may'st be much in my gaskins, but nothing in my
netherstocks.' This is your fault, mistress, to love him that
loves not you. He thinks scorn to do as he's done to. But
if I were as you, I'd cry 'Go by, Hieronimo, go by'.
I'd set mine old debts against my new driblets,   45
And the hare's foot against the goose giblets.

For if ever I sigh when sleep I should take,
Pray God I may lose my maidenhead when I wake.

ROSE Will my love leave me, then, and go to France?

SYBIL I know not that. But I am sure I see him stalk before 50
the soldiers. By my troth, he is a proper man. But he is
proper that proper doth. Let him go snick-up, young
mistress.

ROSE Get thee to London, and learn perfectly
Whether my Lacy go to France or no.   55
Do this, and I will give thee for thy pains
My cambrick apron and my Romish gloves,
My purple stockings and a stomacher.
Say, wilt thou do this, Sybil, for my sake?

SYBIL Will I, quoth 'a? At whose suit? By my troth, yes 60
I'll go—a cambrick apron, gloves, a pair of purple stockings
and a stomacher! I'll sweat in purple, mistress, for you!
I'll take anything that comes, o' God's name! Oh rich—a
cambrick apron! Faith then, have at up tails all. I'll go
jiggy-joggy to London and be here in a trice, young 65
mistress.   *Exit.*

ROSE Do so, good Sybil. Meantime, wretched I
Will sit and sigh for his lost company.   *Exit*

SCENE 3

Tower Street in London. Enter ROWLAND
LACY *like a Dutch shoemaker.*

LACY How many shapes have gods and kings devised
Thereby to compass their desirèd loves?
It is no shame for Rowland Lacy, then,
To clothe his cunning with the Gentle Craft,
That thus disguised I may unknown possess   5
The only happy presence of my Rose.

For her I have forsook my charge in France,
Incurred the King's displeasure, and stirred up
Rough hatred in mine uncle Lincoln's breast.
Oh love, how powerful art thou, that canst change   10
High birth to bareness, and a noble mind
To the mean semblance of a shoemaker!
But thus it must be: for her cruel father,
Hating the single union of our souls,
Hath secretly conveyed my Rose from London,   15
To bar me of her presence. But I trust
Fortune and this disguise will further me
Once more to view her beauty, gain her sight.
Here in Tower Street, with Eyre the shoemaker,
Mean I awhile to work. I know the trade,   20
I learnt it when I was in Wittenberg.
Then cheer thy hoping spirits, be not dismayed:
Thou canst not want, do Fortune what she can.
The Gentle Craft is living for a man.   *Exit.*

SCENE 4

An open yard by Eyre's shop.
Enter EYRE *making himself ready.*

EYRE   Where be these boys, these girls, these drabs, these
scoundrels? They wallow in the fat brewis of my bounty
and lick up the crumbs of my table, yet will not rise to
see my walks cleansed. Come out, you powder-beef
queans! What, Nan! What, Madge Mumblecrust! Come   5
out, you fat midriff swagbelly whores, and sweep me these
kennels, that the noisome stench offend not the nose of
my neighbours. What, Firke, I say! What, Hodge! Open
my shop-windows! What, Firke, I say!

Enter FIRKE.

41

FIRKE   Oh master, is't you that speak bandog and bedlam 10
this morning? I was in a dream, and mused what mad man
was got into the street so early. Have you drunk this
morning that your throat is so clear?

EYRE   Ah, well said, Firke! Well said, Firke! To work, my
fine knave, to work. Wash thy face and thou'lt be more 15
blessed.

FIRKE   Let them wash my face that will eat it. Good master
send for a souse wife if you'll have my face cleaner.

Enter HODGE.

EYRE   Away, sloven! Avaunt, scoundrel! Good-morrow,
Hodge. Good-morrow, my fine foreman.   20

HODGE   Oh master, good-morrow—y'are an early stirrer.
Here's a fair morning—good-morrow, Firke—I could have
slept this hour. Here's a brave day towards!

EYRE   Oh haste to work, my fine foreman. Haste to work.

FIRKE   Master, I am dry as dust to hear my fellow Roger 25
talk of fair weather. Let us pray for good leather, and let
clowns and ploughboys, and those that work in the fields,
pray for brave days. We work in a dry shop: what care I
if it rain?

Enter Eyre's WIFE.

EYRE   How now, Dame Margery? Can you see to rise? 30
Trip and go. Call up the drabs your maids.

WIFE   See to rise? I hope 'tis time enough. 'Tis early
enough for any woman to be seen abroad. I marvel how
many wives in Tower Street are up so soon. God's me,
'tis not noon! Here's a yawling.   35

EYRE   Peace, Margery, peace. Where's Cicely Bumtrinket,
your maid? She has a privy fault—she farts in her sleep.
Call the quean up. If my men want shoe-thread, I'll swinge
her in a stirrup.

FIRKE   Yet that's but a dry beating. Here's still a sign of 40
drought.

Enter LACY *singing, disguised as Hans, a Dutchman.*

LACY   Der was een bore van Gelderland,
   Frolik si byen.
He was als dronk he could nyet stand,
   Upsolce se byen.   45
Tap eens de canneken,
   Drinke schone mannekin.

FIRKE   Master, for my life, yonder's a brother of the Gentle
Craft. If he bear not Saint Hugh's bones, I'll forfeit my
bones. He's some uplandish workman—hire him, good 50
master, that I may learn some gibble-gabble. 'Twill make
us work the faster.

EYRE   Peace, Firke. A hard world! Let him pass, let him
vanish. We have journeymen enough. Peace, my fine
Firke.   55

WIFE   Nay, nay, y'are best to follow your man's counsel.
You shall see what will come on't. We have not men
enough but we must entertain every butterbox, but let that
pass.

HODGE   Dame, 'fore God, if my master follow your counsel, 60
he'll consume little beef. He shall be glad of men an' he can
catch them.

FIRKE   Ay, that he shall.

HODGE   'Fore God, a proper man, and, I warrant, a fine
workman! Master, farewell. Dame, adieu. If such a man 65
as he cannot find work, Hodge is not for you. *Offers to go.*

EYRE   Stay, my fine Hodge.

FIRKE   Faith, an' your foreman go, dame, you must take a
journey to seek a new journeyman. If Roger remove,
Firke follows. If Saint Hugh's bones be not set a-work, I 70

43

may prick mine awl in the walls and go play. Fare ye well, master. Goodbye, dame.

EYRE Tarry, my fine Hodge, my brisk foreman! Stay, Firke. Peace, pudding-broth. By the Lord of Ludgate, I love my men as my life. Peace, you gallimaufry. Hodge, 75 if he want work I'll hire him. One of you, to him...stay, he comes to us.

LACY Goeden dach, meester, ende u vro oak.

FIRKE Nails! If I should speak after him without drinking I should choke. (*to Lacy*) And you, friend Oak, are you of 80 the Gentle Craft?

LACY Yaw, yaw. Ik bin den skomawker.

FIRKE 'Den skomawker', quoth 'a. And heark you, good skomaker: have you all your tools? A good rubbing pin, a good stopper, a good dresser, your four sorts of awls, and 85 your two balls of wax, your paring knife, your hand and thumb-leathers, and good Saint Hugh's bones to smooth up your work?

LACY Yaw, yaw. Be niet vorveard. Ik hab all de dingen voour mak skoes groot and clean. 90

FIRKE Ha ha! Good master, hire him—he'll make me laugh so that I shall work more in mirth than I can in earnest.

EYRE Hear ye, friend. Have ye any skill in the mystery of cordwainers?

LACY Ik weet niet wat yow seg. Ich verstaw you niet. 95

FIRKE Why, thus man (*shows him in mime imitating a shoe-maker at work*). 'Ich verstaw you niet', quoth 'a.

LACY Yaw, yaw, yaw. Ich can dat wel doen.

FIRKE Yaw, yaw. He speaks yawing like a jackdaw that gapes to be fed with cheese curds. Oh, he'll give a villainous 100 pull at a can of double beer. But Hodge and I have the vantage: we must drink first, because we are the eldest journeymen.

EYRE What is thy name?

LACY Hans. Hans Meulter. 105

EYRE Give me thy hand: th'art welcome. Hodge, entertain
him. Firke, bid him welcome. Come, Hans. Run, wife.
Bid your maids, your Trullibubs, make ready my fine
men's breakfasts. To him, Hodge.

HODGE Hans, th'art welcome. Use thyself friendly for we 110
are good fellows. If not, thou shalt be fought with, wert
thou bigger than a giant.

FIRKE Yea, and drunk with, wert thou Gargantua. My
master keeps no cowards, I tell thee. Ho, boy! Bring him
an heel-block: here's a new journeyman. 115

Enter Boy.

LACY Oh, ich wersto you. Ich moet een halve dossen cans
betaelen. Here boy, nempt dis skilling. Tap eens freelick.

Exit Boy.

EYRE Quick, snipper-snapper, away! Firke, scour thy
throat—thou shalt wash it with Castilian liquor. (*to Boy*)
Come, my last of the fives. 120

Enter Boy.

Give me a can. Have to thee, Hans! Here, Hodge! Drink,
you mad Greeks, and work like true Trojans. And pray
for Simon Eyre, the shoemaker. Here, Hans, and th'art
welcome.

FIRKE Lo, dame, you would have lost a good fellow 125
that will teach us to laugh. This beer came hopping in
well.

WIFE Simon, it is almost seven.

EYRE Is't so, Dame Clapper-dudgeon? Is't seven a clock
and my men's breakfast not ready? Trip and go, you 130
soused conger—away! Come, you mad Hyperboreans.

45

Follow me, Hodge. Follow me, Hans. Come after, my
fine Firke. To work, to work a while, and then to break-
fast.   *Exit.*

FIRKE Soft! Yaw, yaw, good Hans. Though my master 135
have no more wit but to call you afore me, I am not so
foolish to go behind you, I being the elder journeyman.

                                                              Exeunt.

ACT II

SCENE I

A field near Old Ford. Hallooing within. Enter
WARNER and HAMMON, dressed as hunters.

HAM. Cousin, beat every brake—the game's not far.
This way with wingèd feet he fled from death,
Whilst the pursuing hounds scenting his steps
Find out his high way to destruction.
Besides, the miller's boy told me even now   5
He saw him take soil, and he halloed him,
Affirming him so embossed
That long he could not hold.
WAR.   If it be so,
'Tis best we trace these meadows by Old Ford.

A noise of hunters within. Enter a Boy.

HAM. How now, boy? Where's the deer? Speak, saw'st 10
thou him?
BOY Oh yea, I saw him leap through a hedge and then over
a ditch. Then at my Lord Mayor's pale, over he skipped me
and in he went me, and 'Holla' the hunters cried and
'There boy, there boy', but there he is, a'mine honesty.   15
HAM. Boy, God amercy! Cousin, let's away.
I hope we shall find better sport today.   *Exeunt.*

SCENE 2

Another part of the field, nearer the house.
Hunting within. Enter ROSE and SYBIL.

ROSE Why, Sybil, wilt thou prove a forester?
SYBIL Upon some, no! Forester, go by! No, faith, mistress,

the deer came running into the barn through the orchard
and over the pale. I wot well I looked as pale as a new
cheese to see him. But 'Whip!' says Goodman Pinclose, 5
up with his flail and our Nick with a prong. And down he
fell, and they upon him, and I upon them. By my troth,
we had such sport, and in the end we ended him. His
throat we cut, flayed him, unhorned him, and my Lord
Mayor shall eat of him anon when he comes. 10

Horns sound within.

ROSE Hark, hark, the hunters come! Y'are best take heed:
They'll have a saying to you for this deed.

Enter HAMMON, WARNER, *huntsmen and boy.*

HAM. God save you, fair ladies.
SYBIL   Ladies! Oh, gross!
WAR. Came not a buck this way?
ROSE   No, but two does.
HAM. And which way went they? Faith, we'll hunt at those. 15
SYBIL At those? Upon some, no! When, can you tell?
WAR. Upon some, ay!
SYBIL   Good Lord!
WAR.   Wounds! Then farewell.
HAM. Boy, which way went he?
BOY   This way, sir, he ran.
HAM. This way he ran indeed. Fair mistress Rose,
  Our game was lately in your orchard seen. 20
WAR. Can you advise which way he took his flight?
SYBIL Follow your nose—his horns will guide you right.
WAR. Th'art a mad wench.
SYBIL   Oh rich!
ROSE   Trust me, not I.
  It is not like, the wild forest deer

48

Would come so near to places of resort.   25
    You are deceived: he fled some other way.
WAR. Which way, my sugar-candy, can you show?
SYBIL Come up, good honey-sops. Upon some, no!
ROSE Why do you stay, and not pursue your game?
SYBIL I'll hold my life their hunting nags be lame.   30
HAM. A deer more dear is found within this place.
ROSE But not the deer, sir, which you had in chase.
HAM. I chased the deer, but this dear chaseth me.
ROSE The strangest hunting that ever I see.
    But where's your park?   *She offers to go away*
HAM.   'Tis here—oh stay!   35
ROSE Impale me and then I will not stray.
WAR. (*to Sybil*) They wrangle, wench. We are more kind
    than they.
SYBIL What kind of hart is that, dear heart, you seek?
WAR. A hart, dear heart.
SYBIL   Who ever saw the like?
ROSE To lose your heart, is't possible you can?   40
HAM. My heart is lost.
ROSE   Alack, good gentleman!
HAM. This poor lost heart would I wish you might find.
ROSE You, by such luck, might prove your hart a hind.
HAM. Why, Luck had horns, so have I heard some say.
ROSE Now God, an't be His will, send Luck into your way. 45

Enter LORD MAYOR *and servants.*

L.MA. What, Master Hammon! Welcome to Old Ford.
SYBIL (*to Warner*) God's pittikins, hands off, sir! Here's my
    Lord.
L.MA. I hear you had ill luck and lost your game.
HAM. 'Tis true, my Lord.
L.MA.   I am sorry for the same.

What gentleman is this?

HAM.   My brother-in-law.   50

L.MA. Y'are welcome both. Sith Fortune offers you

   Into my hands, you shall not part from hence

   Until you have refreshed your wearied limbs.

   Go, Sybil, cover the board. You shall be guest

   To no good cheer but even a hunter's feast.   55

HAM. I thank your Lordship. (*aside to Warner*) Cousin, on my life,

   For our lost venison, I shall find a wife.   *Exeunt.*

L.MA. In, gentlemen. I'll not be absent long.

   This Hammon is a proper gentleman,

   A citizen by birth, fairly allied.   60

   How fit an husband were he for my girl!

   Well, I will in and do the best I can

   To match my daughter to this gentleman.   *Exit.*

SCENE 3

Eyre's shop. Enter LACY (*as Hans*),
SKIPPER, HODGE *and* FIRKE.

SKIP. Ick sal yow wat seggen, Hans: dis skip dat comen from
   Candy is al vol, by Got's sacrament, van sugar, civet,
   almonds, cambrick and all dingen—towsand, towsand ding.
   Nempt it, Hans. Nempt it vor u meester. Daer be bills
   van laden. Your meester, Simon Eyre, sal hae good copen. 5
   Wat seggen you, Hans?

FIRKE Wat seggen de reggen de copen, slopen. Laugh,
   Hodge, laugh!

LACY Mine liever broder, Firke, bringt meester Eyre tot
   den sign van swannekin. Daer sal yow find dis skipper end 10
   me. Wat seggen yow, broder Firke? Doot it, Hodge.
   Come, skipper.   *Exeunt Lacy and Skipper.*

FIRKE 'Bring him', quoth you? Here's no knavery, to

bring my master to buy a ship worth the lading of two or three hundred thousand pounds! Alas, that's nothing! A 15 trifle, a bable, Hodge.

HODGE The truth is, Firke, that the merchant owner of the ship dares not show his head, and therefore this skipper that deals for him, for the love he bears to Hans, offers my master Eyre a bargain in the commodities. He shall have 20 a reasonable day of payment. He may sell the wares by that time and be an huge gainer himself.

FIRKE Yea, but can my fellow Hans lend my master twenty porpentines as an earnest penny?

HODGE Portuguese, thou wouldst say. Here they be, Firke. 25 Hark, they jingle in my pocket like Saint Mary Overy's bells.

Enter EYRE, *his* WIFE *and a Boy.*

FIRKE Mum! Here comes my dame and my master. She'll scold, on my life, for loitering this Monday. But all's one: let them all say what they can, Monday's our holiday. 30

WIFE You sing, Sir Sauce, but I beshrew your heart:
I fear for this your singing we shall smart.

FIRKE Smart for me, dame? Why, dame, why?

HODGE Master, I hope you'll not suffer my dame to take down your journeyman. 35

FIRKE If she take me down, I'll take her up. Yea, and take her down too—a button-hole lower.

EYRE Peace, Firke. Not I, Hodge, by the life of Pharaoh. By the Lord of Ludgate, by this beard, every hair whereof I value at a king's ransom: she shall not meddle with you. 40 Peace, you bumbast-cotton-candle quean! Away, Queen of Clubs! Quarrel not with my men, with me and my fine Firke. I'll firke you if you do.

WIFE Yea, yea, man. You may use me as you please. But let that pass. 45

51   4-2

EYRE Let it pass, let it vanish away. Peace! Am I not Simon
Eyre? Are not these my brave men? Brave shoemakers, all
gentlemen of the Gentle Craft. Prince am I none, yet am
I nobly born, as being the sole son of a shoemaker. Away,
rubbish! Vanish! Melt like kitchen stuff.   50

WIFE Yea, yea, 'tis well. I must be called rubbish, kitchen
stuff, for a sort of knaves.   *Weeps.*

FIRKE Nay, dame. You shall not weep and wail in woe for
me. Master, I'll stay no longer. Here's a venentory of my
shop tools. Adieu, master. Hodge, farewell.   55

HODGE Nay, stay, Firke. Thou shalt not go alone.

WIFE I pray let them go. There be more maids than Mawkin,
more men than Hodge, and more fools than Firke.

FIRKE Fools! Nails! If I tarry now I would my guts might
be turned to shoe-thread!   60

HODGE And if I stay, I pray God I may be turned to a Turk
and set in Finsbury for boys to shoot at. Come, Firke.

EYRE Stay, my fine knaves! You arms of my trade, you
pillars of my profession. What, shall a tittle-tattle's words
make you forsake Simon Eyre? (*to his Wife*) Avaunt, 65
kitchen-stuff! Rip, you brown-bread tannikin, out of my
sight! Move me not! Have I not ta'en you from selling
tripes in Eastcheap and set you in my shop and made you
hail-fellow with Simon Eyre the shoemaker? And now do
you deal thus with my journeymen? Look, you powder- 70
beef quean, on the face of Hodge. Here's a face for a
lord.

FIRKE And here's a face for any lady in Christendom.

EYRE Rip, you chitterling! Avaunt, boy: bid the tapster
of the Bore's Head fill me a dozen cans of beer for my 75
journeymen.

FIRKE A dozen cans? O brave! Hodge, now I'll stay.

EYRE (*aside to the boy*) An the knave fills any more than two,

he pays for them. (*exit Boy*) (*aloud*) A dozen cans of beer
for my journeymen!   80

      Enter Boy with two cans, and exit.

Here, you mad Mesopotamians. Wash your livers with this
liquor. Where be the odd ten? No more, Madge, no more.
Well said. Drink and to work. What work dost thou,
Hodge? What work?

HODGE I am making a pair of shoes for Sybil, my Lord's 85
maid. I deal with her.

EYRE Sybil? Fie, defile not thy fine workmanly fingers with
the feet of kitchen-stuff and basting ladles! Ladies of the court,
fine ladies, my lads! Commit *their* feet to our apparelling.
Put gross work to Firke. Yark and seam, yark and seam! 90

FIRKE For yarking and seaming let me alone, an I come to't.

HODGE Well, master, all this is from the bias. Do you
remember the ship my fellow Hans told you of? The skipper
and he are both drinking at the Swan. Here be the Portuguese to give earnest. If you go through with it you cannot 95
choose but be a lord at least.

FIRKE Nay, dame, if my master prove not a lord and you
a lady hang me.

WIFE Yea, like enough, if you loiter and tipple thus.

FIRKE Tipple, dame? No, we have been bargaining with 100
Skellum Skanderbag. Can you Dutch spreaken for a ship
of silk Cyprus, laden with sugar-candy?

EYRE Peace, Firke. Silence, tittle-tattle. Hodge, I'll go
through with it. Here's a ring, and I have sent for a
guarded gown and a damask cassock.   105

  Enter the Boy with a velvet coat and an Alderman's gown.

See where it comes. Look here, Maggy. (*Eyre puts it on*)
Help me, Firke. Apparel me, Hodge. Silk and satin, you
mad Philistines! Silk and satin!

FIRKE Ha ha! My master will be as proud as a dog in a
doublet, all in beaten damask and velvet.   110

EYRE Softly, Firke, for rearing of the nap and wearing
threadbare my garments. How dost thou like me, Firke?
How do I look, my fine Hodge?

HODGE Why, now you look like yourself, master. I warrant
you, there's few in the City but will give you the wall and   115
come upon you with the Right-Worshipful.

FIRKE Nails! My master looks like a threadbare cloak new
turned and dressed. Lord, Lord, to see what good raiment
doth! Dame, dame, are you not enamoured?

EYRE How say'st thou, Maggy? Am I not brisk? Am I not   120
fine?

WIFE Fine? By my troth, sweetheart, very fine. By my
troth, I never liked thee so well in my life, sweet heart. But
let that pass. I warrant there be many women in the City
have not such handsome husbands, but only for their   125
apparel. But let that pass too.

Enter LACY *as Hans, and* SKIPPER.

LACY Godden day, mester. Dis be de skipper dat heb de
skip van merchandice. De commodity ben good. Nempt
it, master. Nempt it.

EYRE God-a-mercy, Hans. Welcome, skipper. Where lies   130
this ship of merchandise?

SKIP. De skip ben in revere. Dor be van sugar, civet, almonds, cambrick, and a towsand towsand tings, Got's
sacrament! Nempt it, mester. Yo sall heb good copen.

FIRKE To him, master. Oh sweet master! Oh sweet wares: 135
prunes, almonds, sugar-candy, carrot roots, turnips! Oh
brave fatting meat! Let not a man buy a nutmeg but yourself.

EYRE Peace, Firke. Come, skipper, I'll go aboard with you.
Hans, have you made him drink?

SKIP. Yaw, yaw, ic hab veal gedrunk.   140

EYRE   Come, Hans, follow me. Skipper, thou shalt have
my countenance in the City.

Exeunt Eyre, Skipper and Lacy.

FIRKE   'Yaw, heb veal ge drunk', quoth-a. They may well be
called butter-boxes when they drink fat veal, and thick beer
too. But come, dame, I hope you'll chide us no more.   145

WIFE   No faith, Firke. No perdy, Hodge. I do feel honour
creep upon me. And, which is more, a certain rising in
my flesh, but let that pass.

FIRKE   Rising in your flesh do you feel, say you? Ay, you
may be with child. But why should not my master feel a 150
rising in his flesh, having a gown and a gold ring on. But
you are such a shrew, you'll soon pull him down.

WIFE   Ha ha, prithee peace. Thou mak'st my worship laugh.
But let that pass. Come, I'll go in. Hodge, prithee go
before me. Firke, follow me.   155

FIRKE   Firke doth follow. Hodge, pass out in state.

Exeunt.

SCENE 4

Enter LINCOLN *and* DODGER.

LINC. How now, good Dodger? What's the news in
France?

DODGER   My Lord, upon the eighteenth day of May
The French and English were prepared to fight.
Each side with eager fury gave the sign
Of a most hot encounter. Five long hours   5
Both armies fought together. At the length,
The lot of victory fell on our sides.
Twelve thousand of the Frenchmen that day died;
Four thousand English, and no man of name
But Captain Hyam and young Ardington.   10

55

LINC. Two gallant gentlemen, I knew them well.
 But, Dodger, prithee tell me, in this fight
 How did my cousin Lacy bear himself?

DODGER My Lord, your cousin Lacy was not there.

LINC. Not there?

DODGER   No, my good Lord.

LINC.   Sure thou mistakest. 15
 I saw him shipped, and a thousand eyes beside
 Were witnesses of the farewells which he gave
 When I with weeping eyes bid him adieu.
 Dodger, take heed.

DODGER   My Lord, I am advised
 That what I spake is true. To prove it so, 20
 His cousin Askew, that supplied his place,
 Sent me for him from France, that secretly
 He might convey himself hither.

LINC.   Is't even so?
 Dares he so carelessly venture his life
 Upon the indignation of a King? 25
 Hath he despis'd my love and spurn'd those favours
 Which I with prodigal hand pour'd on his head?
 He shall repent his rashness with his soul.
 Since of my love he makes no estimate,
 I'll make him wish he had not known my hate. 30
 Thou hast no other news?

DODGER   None else, my Lord.

LINC. None worse I know thou hast. Procure the King
 To crown his giddy brows with ample honours!
 Send him chief Colonel, and all my hope
 Thus to be dashed! But 'tis in vain to grieve. 35
 One evil cannot a worse relieve.
 Upon my life, I have found out his plot!
 That old dog Love that fawn'd upon him so,

Love to that puling girl, his fair-cheeked Rose,
The Lord Mayor's daughter, hath distracted him;   40
And in the fire of that love's lunacy
Hath he burnt up himself, consum'd his credit,
Lost the King's love, yea, and I fear, his life:
Only to get a wanton to his wife!
Dodger, it is so.
DODGER   I fear so, my good Lord.   45
LINC. It is so. Nay, sure it cannot be!
  I am at my wit's end. Dodger—
DODGER   Yea, my Lord.
LINC. Thou art acquainted with my nephew's haunts.
  Spend this gold for thy pains. Go, seek him out.
  Watch at my Lord Mayor's. There, if he live,   50
  Dodger, thou shalt be sure to meet with him.
  Prithee, be diligent. Lacy, thy name
  Lived once in honour, now dead in shame.
  (*to Dodger*) Be circumspect.   *Exit.*
DODGER I warrant you, my Lord.   *Exit.*

ACT III

SCENE I

A room in the Lord Mayor's house at London.
Enter LORD MAYOR *and* MASTER SCOTT.

L. MA. Good Master Scott, I have been bold with you
To be a witness to a wedding knot
Betwixt young Master Hammon and my daughter.
Oh, stand aside—see where the lovers come.

Enter HAMMON *and* ROSE.

ROSE Can it be possible you love me so?   5
No, no! Within these eyeballs I espy
Apparent likelihoods of flattery.
Pray now, let go my hand.
HAM.   Sweet Mistress Rose,
Misconstrue not my words, nor misconceive
Of my affection, whose devoted soul   10
Swears that I love thee dearer than my heart.
ROSE As dear as your own heart? I judge it right:
Men love their hearts best when th'are out of sight.
HAM. I love you, by this hand!
ROSE   Yet hands off now!
If flesh be frail, how weak and frail's your vow!   15
HAM. Then by my life I swear.
ROSE   Then do not brawl.
One quarrel loseth wife and life and all.
Is not your meaning thus?
HAM.   In faith, you jest.
ROSE Love loves to sport; therefore leave love, y'are
best.

L.MA. What? Square they, Master Scott?

SCOTT   Sir, never doubt,   20
   Lovers are quickly in and quickly out.

HAM. Sweet Rose, be not so strange in fancying me.
   Nay, never turn aside! Shun not my sight.
   I am not grown so fond to fond my love
   On any that shall quit it with disdain.   25
   If you will love me, so; if not, farewell.

L.MA. Why, how now, lovers? Are you both agreed?

HAM. Yes, faith, my Lord.

L.MA.   'Tis well. Give me your hand.
   Give me yours, daughter. How now? Both pull back?
   What means this, girl?

ROSE   I mean to live a maid.   30

HAM. (*aside*) But not to die one—pause ere that be said.

L.MA. Will you still cross me? Still be obstinate?

HAM. Nay, chide her not, my Lord, for doing well.
   If she can live an happy virgin's life
   'Tis far more blessed than to be a wife.   35

ROSE Say, sir, I cannot. I have made a vow:
   Whoever be my husband, 'tis not you.

L.MA. Your tongue is quick. But, Master Hammond, know
   I bade you welcome to another end.

HAM. What? Would you have me pule and pine and pray   40
   With 'lovely lady, mistress of my heart',
   'Pardon your servant' and the rhymer play,
   Railing on Cupid and his tyrant dart?
   Or shall I undertake some martial spoil,
   Wearing your glove at tourney and at tilt,   45
   And tell how many gallants I unhorsed?
   Sweet, will this pleasure you?

ROSE   Yea. When wilt begin?
   What? Love-rhymes, man? Fie on that deadly sin.

L.MA. If you will have her, I'll make her agree.

HAM. Enforcèd love is worse than hate to me.   50
There is a wench keeps shop in the Old Change.
To her will I. It is not wealth I seek:
I have enough, and will prefer her love
Before the world. My good Lord Mayor, adieu.
Old love for me: I have no luck with new.   *Exit.* 55

L.MA. (*to Rose*) Now, mammet! You have well behaved
yourself!
But you shall curse your coyness if I live.
Who's within there? See you convey your mistress
Straight to th' Old Ford. I'll keep you straight enough.
Fore God, I would have sworn the puling girl   60
Would willingly accepted Hammon's love.
But banish him my thoughts! Go, minion, in! *Exit Rose.*
Now, tell me, Master Scott, would you have thought
That Master Simon Eyre, the shoemaker,
Had been of wealth to buy such merchandise?   65

SCOTT 'Twas well, my Lord, your Honour and myself
Grew partners with him. For your bills of lading
Show that Eyre's gains in one commodity
Rise at the least to full three thousand pound,
Besides like gain in other merchandise.   70

L.MA. Well, he shall spend some of his thousands now,
For I have sent for him to the Guildhall.

Enter EYRE.

See where he comes. Good morrow, Master Eyre.

EYRE Poor Simon Eyre, my Lord, your shoemaker.

L.MA. Well, well, it likes yourself to term you so.   75

Enter DODGER.

Now, Master Dodger, what's the news with you?

DODGER I'll gladly speak in private to your Honour.

L.MA. You shall, you shall. Master Eyre and Master Scott,
　I have some business with this gentleman.
　I pray you, let me entreat you to walk before　　　　　80
　To the Guildhall. I'll follow presently.
　Master Eyre, I hope ere noon to call you Sheriff.
EYRE I would not care, my Lord, if you might call me King
　of Spain. Come, Master Scott.　　*Exeunt Eyre and Scott.*
L.MA. Now, Master Dodger, what's the news you bring?　85
DODGER The Earl of Lincoln by me greets your Lordship
　And earnestly requests you, if you can,
　Inform him where his nephew Lacy keeps.
L.MA. Is not his nephew Lacy now in France?
DODGER No, I assure your Lordship, but disguised　　　90
　Lurks here in London.
L.MA.　　　　　　London? Is't even so?
　It may be, but upon my faith and soul
　I know not where he lives or whether he lives.
　So tell my Lord of Lincoln. Lurk in London?
　Well, Master Dodger, you perhaps may start him.　95
　Be but the means to rid him into France,
　I'll give you a dozen angels for your pains:
　So much I love his Honour, hate his nephew,
　And prithee so inform thy Lord from me.
DODGER I take my leave.
L.MA.　　　　　　Farewell, good Master Dodger.　100
　　　　　　　　　　　　　Exit Dodger.
　Lacy in London? I dare pawn my life
　My daughter knows thereof and for that cause
　Denied young Master Hammon in his love.
　Well, I am glad I sent her to Old Ford.
　God's Lord, 'tis late! To Guildhall I must hie.　105
　I know my brethren stay my company.　　　*Exit.*

Act III, Scene 2

SCENE 2

A room in Eyre's house. Enter FIRKE, *Eyre's* WIFE,
LACY *as Hans, and* ROGER.

WIFE   Thou goest too fast for me, Roger. Oh, Firke!

FIRKE   Ay, forsooth.

WIFE   I pray thee, run, do you hear, run to Guildhall and
learn if my husband Master Eyre will take that worshipful
vocation of Master Sheriff upon him. Hie thee, good Firke.   5

FIRKE   Take it? Well, I go. An he should *not* take it, Firke
swears to forswear him. Yes, forsooth, I go to Guildhall.

WIFE   Nay, when? Thou art too compendious and tedious.

FIRKE   Oh rare! Your Excellence is full of eloquence. *(aside)*
How like a new cart-wheel my dame speaks! And she   10
looks like an old, musty ale-bottle going to scalding.

WIFE   Nay, when? Thou wilt make me melancholy.

FIRKE   God forbid your Worship should fall into that
humour. I run.   *Exit.*

WIFE   Let me see now, Roger and Hans.   15

HODGE   Ay, forsooth, dame—mistress, I should say but the
old term so sticks to the roof of my mouth I can hardly lick
it off.

WIFE   Even what thou wilt, good Roger. Dame is a fair
name for any honest Christian, but let that pass. How dost   20
thou, Hans?

LACY   Me tank you, vro.

WIFE   Well, Hans and Roger, you see God hath blest your
master, and, perdy, if ever he comes to be Master Sheriff
of London (as we are all mortal) you shall see I will have   25
some odd thing or other in a corner for you. I will not be
your back friend. But let that pass. Hans, pray thee tie my
shoe.

LACY   Yaw, ic sall, vro.

62

WIFE   Roger, thou knowest the length of my foot. As it is 30
none of the biggest, so I thank God it is handsome enough.
Prithee, let me have a pair of shoes made. Cork, good
Roger, wooden heel too.

HODGE   You shall.

WIFE   Art thou acquainted with never a farthingale maker, 35
nor a French-hood maker. I must enlarge my bum. Ha ha!
How shall I look in a hood, I wonder? Perdy, oddly I
think.

HODGE   (*aside*) As a cat out of a pillory. (*to her*) Very well,
I warrant you, mistress.   40

WIFE   Indeed, all flesh is grass. And, Roger, canst thou tell
where I may buy a good hair?

HODGE   Yea, forsooth, at the poulterer's in Gracious Street.

WIFE   Thou art an ungracious wag, perdy. I mean a false
hair for my periwig.   45

HODGE   Why, mistress, the next time I cut my beard you
shall have the shavings of it—but they are all true hairs.

WIFE   It is very hot. I must get me a fan or else a mask.

HODGE   (*aside*) So you had need, to hide your wrinkled face.

WIFE   Fie upon it! How costly this world's calling is, perdy. 50
But that is one of the wonderful works of God: I would not
deal with it. Is not Firke come yet? Hans, be not so sad:
let it pass and vanish, as my husband's Worship says.

LACY   Ic bin vrolick. Lot see yow so.

HODGE   Mistress, will you drink a pipe of tobacco?   55

WIFE   Oh fie upon it, Roger! Perdy, these filthy tobacco
pipes are the most idle, slavering bables that ever I felt.
Out upon it! God bless us, men look not like men that use
them.
                    Enter RAFE, *being lame.*

HODGE   What! Fellow Rafe? Mistress, look here—Jane's 60
husband! Why, how now—lame? Hans, make much of him.

He's a brother of our trade, a good workman and a tall
soldier.

LACY   You be welcome, broder.

WIFE   Perdy, I knew him not. How doest thou, good Rafe? 65
I am glad to see thee well.

RAFE   I would God you saw me, Dame, as well
As when I went from London into France.

WIFE   Trust me, I am sorry, Rafe, to see thee impotent.
Lord, how the wars have made him sunburnt! The left leg 70
is not well. 'Twas a fair gift of God the infirmity took not
hold a little higher, considering thou camest from France.
But let that pass.

RAFE   I am glad to see you well, and I rejoice
To hear that God hath blest my master so   75
Since my departure.

WIFE   Yea, truly, Rafe, I thank my maker. But let that pass.

HODGE   And, sirrah Rafe, what news? What news in France?

RAFE   Tell me, good Roger, first, what news in England?
How does my Jane? When didst thou see my wife?   80
Where lives my poor heart? She'll be poor indeed
Now I want limbs to get whereon to feed.

HODGE   Limbs? Hast thou not hands, man? Thou shalt never
see a shoemaker want bread, though he have but three
fingers on a hand.   85

RAFE   Yet all this while I hear not of my Jane.

WIFE   Oh, Rafe! Your wife, perdy, we know not what's
become of her. She was here awhile, and because she was
married grew more stately than became her. I checked her,
and so forth. Away she flung. Never returned, nor said 90
Bye nor Bah. And, Rafe, you know 'ka me, ka thee'. And
so, as I tell ye. Roger, is not Firke come yet?

HODGE   No, forsooth.

WIFE   And so, indeed, we heard not of her, but I hear she

lives in London (but let that pass). If she had wanted, she 95
might have opened her case to me or my husband, or to
any of my men. I am sure there's not any of them, perdy,
but would have done her good to his power. Hans, look
if Firke be come.

LACY   Yaw, ic sal, vro.   *Exit.* 100

WIFE   And so as I said. But, Rafe, why dost thou weep?
Thou knowest that naked we came out of our mother's
womb, and naked we must return. And therefore thank
God for all things.

HODGE   No, faith. Jane is a stranger here, but, Rafe, pull up 105
a good heart—I know thou hast one. Thy wife, man, is in
London. One told me he saw her a while ago, very brave
and neat. We'll ferret her out, an' London hold her.

WIFE   Alas, poor soul. He's overcome with sorrow. He
does but as I do—weep for the loss of any good thing. But, 110
Rafe, get thee in. Call for some meat and drink. Thou shalt
find me worshipful towards thee.

RAFE   I thank you, dame. Since I want limbs and lands,
I'll to God, my good friends, and these my hands.   *Exit.*

Enter LACY (*as Hans*), *and* FIRKE *running.*

FIRKE   Run, good Hans. Oh, Hodge! Oh, Mistress! 115
Hodge, heave up thine ears. Mistress, smug up your looks.
On with your best apparel. My master is chosen! My
master is called, nay, condemned by the cry of the country,
to be Sheriff of the City for this famous year now to come
and time now being. A great many men in black gowns 120
were asked for their voices and their hands, and my master
had all their fists about his ears presently. And they cried
'Ay, ay, ay, ay', and so I came away.
Wherefore, without all other grieve,
I do salute you Mistress Shrieve.   125

5   65   SD

LACY   Yaw. My mester is de groot man, de shrieve.

HODGE   Did I not tell you, Mistress? Now I may boldly say 'Good morrow to your Worship'.

WIFE   Good morrow, good Roger. I thank you, my good people all. Firke, hold up thy hand. Here's a three-penny 130 piece for thy tidings.

FIRKE   'Tis but three-half-pence, I think. Yes, 'tis three-pence: I smell the rose.

HODGE   But, Mistress, be ruled by me, and do not speak so pulingly.   135

FIRKE   'Tis her Worship speaks so and not she. No faith, Mistress, speak me in the old key—'To it, Firke! There, good Firke! Ply your business, Hodge!' 'Hodge!' with a full mouth! 'I'll fill your bellies with good cheer till they cry twang.'   140

Enter SIMON EYRE *wearing a gold chain.*

LACY   See, mine liever broder! Here compt my meester.

WIFE   Welcome home, Master Shrieve! I pray God continue you in health and wealth.

EYRE   See here, my Maggy. A chain, a gold chain for Simon Eyre. I shall make thee a lady. Here's a French hood for 145 thee—on with it, on with it! Dress thy brows with this flap of a shoulder of mutton to make thee look lovely. Where be my fine men? Roger, I'll make over my shop and tools to thee. Firke, thou shalt be the foreman. Hans, thou shalt have an hundred for twenty. Be as mad knaves 150 as your master Sim Eyre hath been, and you shall live to be Sheriffs of London. How dost thou like me, Margery? Prince am I none, yet am I princely born. Firke, Hodge and Hans!

ALL THREE   Ay forsooth. What says your Worship Master 155 Sheriff?

66

EYRE Worship and honour, you Babylonian knaves, for the
Gentle Craft! But I forget myself. I am bidden by my
Lord Mayor to dinner at Old Ford. He's gone before, I
must after. Come, Madge, on with your trinkets. Now, 160
my true Trojans, my fine Firke, my dapper Hodge, my
honest Hans. Some device, some odd crotchets, some
morris or such like, for the honour of the gentle shoemakers.
Meet me at Old Ford. You know my mind.
Come, Madge, away!   165
Shut up the shop, knaves, and make holiday!
                               Exeunt Simon and Mistress Eyre.
FIRKE Oh rare, oh brave! Come, Hodge. Follow me, Hans.
We'll be with them for a morris dance.   *Exeunt.*

SCENE 3

A room at Old Ford. Enter LORD MAYOR, EYRE,
his WIFE *in a French hood,* ROSE, SYBIL *and other servants.*

L.MA. Trust me, you are as welcome to Old Ford
As I myself.
WIFE Truly, I thank your Lordship.
L.MA. Would our bad cheer were worth the thanks you
give.   5
EYRE Good cheer, my Lord Mayor, a fine cheer! A fine
house, fine walls, all fine and neat.
L.MA. Now, by my troth, I'll tell thee, Master Eyre,
It does me good and all my brethren
That such a madcap fellow as thyself   10
Is entered into our society.
WIFE Ay, but, my Lord, he must learn now to put on
gravity.
EYRE Peace, Maggy. A fig for gravity! When I go to
Guildhall in my scarlet gown, I'll look as demurely as a 15

saint, and speak as gravely as a Justice of the Peace. But
now I am here at Old Ford, at my good Lord Mayor's
house, let it go by! Vanish, Maggy! I'll be merry. Away
with flip-flap, these fooleries, these gulleries. What, honey?
Prince am I none, yet am I princely born. What says my 20
Lord Mayor?

L.MA. Ha, ha, ha! I had rather than a thousand pound
I had an heart but half so light as yours.

EYRE Why, what should I do, my Lord? A pound of care
pays not a dram of debt. Hum, let's be merry whiles we 25
are young. Old age, sack and sugar will steal upon us ere
we be aware.

L.MA. It's well done. Mistress Eyre, pray give good counsel
to my daughter.

WIFE I hope Mistress Rose will have the grace to take 30
nothing that's bad.

L.MA. Pray God she do, for, i'faith, Mistress Eyre,
I would bestow upon that peevish girl
A thousand marks more than I mean to give her
Upon condition she'd be ruled by me.   35
The ape still crosseth me. There came of late
A proper gentleman, of fair revenues,
Whom gladly I would call son-in-law.
But my fine cockney would have none of him.
You'll prove a cockscomb for it ere you die: 40
A courtier, or no man, must please your eye.

EYRE Be ruled, sweet Rose. Th'art ripe for a man. Marry
not with a boy, that has no more hair on his face than thou
hast on thy cheeks. A courtier! Wash, go by! Stand not
upon pishery pashery. Those silken fellows are but painted 45
images—outsides. Outsides, Rose! Their inner linings are
torn. No, my fine mouse: marry me with a gentleman
grocer, like my Lord Mayor, your father. A grocer is a

sweet trade: plums, plums! Had I a son or daughter should
marry out of the generation and blood of shoemakers, he 50
should pack. What? The Gentle Trade is a living for a man
through Europe, through the world.

A noise within of a tabor and a pipe.

L.MA. What noise is this?
EYRE Oh my Lord Mayor, a crew of fellows that for love
of your Honour are come hither with a morris dance. 55
Come in, my Mesopotamians, cheerly!

Enter HODGE, LACY (*as Hans*), RAFE, FIRKE *and other shoemakers in a morris. After a little dancing, the Lord Mayor
speaks.*

L.MA. Master Eyre, are all these shoemakers?
EYRE All cordwainers, my good Lord Mayor.
ROSE (*aside*) How like my Lacy looks yond shoemaker.
LACY (*aside*) Oh that I durst but speak unto my love!   60
L.MA. Sybil, go fetch some wine to make these drink.
You are all welcome.
ALL We thank your Lordship.

Rose takes a cup of wine and goes to Hans.

ROSE For his sake whose fair shape thou representst,
Good friend, I drink to thee.   65
LACY Ic be dank, good frister.
WIFE I see, Mistress Rose, you do not want judgement. You
have drunk to the properest man I keep.
FIRKE Here be some have done their parts to be as proper
as he.   70
L.MA. Well, urgent business calls me back to London.
Good fellows, first go in and taste our cheer,
And to make merry as you homeward go,
Spend these two angels in beer at Stratford Bow.

EYRE   To these two, my mad lads, Sim Eyre adds another. 75
Then cheerly, Firke. Tickle it, Hans. And all for the honour
of shoemakers.   *All go dancing out.*

L.MA. Come, Master Eyre, let's have your company.
                                    Exeunt Eyre and Lord Mayor.

ROSE   Sybil, what shall I do?

SYBIL Why, what's the matter?   80

ROSE   That Hans the shoemaker is my love Lacy,
Disguised in that attire to find me out.
How should I find the means to speak with him?

SYBIL What, Mistress, never fear. I dare venture my
maidenhead to nothing—and that's great odds—that Hans 85
the Dutchman when we come to London shall not only
see and speak with you, but in spite of all your father's
policies, steal you away and marry you. Will not this
please you?

ROSE   Do this, and ever be assured of my love.   90

SYBIL Away then, and follow your father to London, lest
your absence cause him to suspect something.
Tomorrow, if my counsel be obeyed,
I'll bind you prentice to the Gentle Trade.   *Exeunt.*

SCENE 4

A seamster's shop in London. Enter JANE *working,*
and HAMMON *muffled at another door. He stands aloof.*

HAM.   Yonder's the shop, and there my fair love sits.
She's fair and lovely, but she is not mine.
O, would she were! Thrice have I courted her.
Thrice hath my hand been moistened with her hand,
Whilst my poor famished eyes do feed on that   5
Which made them famished. I am infortunate:
I still love one, yet nobody loves me.

I muse in other men what women see
That I so want! Fine Mistress Rose was coy,
And this too curious—oh no! She is chaste,   10
And, for she thinks me wanton, she denies
To cheer my cold heart with her sunny eyes.
How prettily she works. Oh, pretty hand!
Oh, happy work! It doth me good to stand
Unseen to see her. Thus I oft have stood   15
In frosty evenings, a light burning by her,
Enduring biting cold, only to eye her.
One only look hath seem'd as rich to me
As a king's crown, such is love's lunacy.
Muffled, I'll pass along, and by that try   20
Whether she knows me.

JANE   Sir, what is't you buy?
What is't you lack? Calico? Or lawn?
Fine cambrick shirts, or bands? What will you buy?

HAM. *(aside)* That which thou wilt not sell. Faith, yet I'll try.
(to her) How do you sell this handkercher?

JANE   Good cheap.   25

HAM. And how these ruffs?

JANE   Cheap too.

HAM.   And how this band?

JANE Cheap too.

HAM.   All cheap! How sell you then this hand?

JANE My hands are not to be sold.

HAM.   To be given then.
Nay, faith, I come to buy.

JANE   But none knows when.

HAM. Good sweet, leave work a little while. Let's play.   30

JANE I cannot live by keeping holiday.

HAM. I'll pay you for the time which shall be lost.

JANE With me you shall not be at so much cost.

HAM. Look how you wound this cloth. So you wound me.

JANE It may be so.

HAM.   'Tis so.

JANE   What remedy?   35

HAM. Nay, faith, you are too coy.

JANE   Let go my hand.

HAM. I will do any task at your command.
  I would let go this beauty, were I not
  Enjoined to disobey you by a power
  That controls kings. I love you.

JANE   So: now part.   40

HAM. With hands I may, but never with my heart.
  In faith, I love you.

JANE   I believe you do.

HAM. Shall a true love in me breed hate in you?

JANE I hate you not.

HAM.   Then you must love.

JANE   I do.
  What? Are you better now? I love not you.   45

HAM. All this I hope is but a woman's fray
  That means 'Come to me' when she cries 'Away'.
  In earnest, mistress, I do not jest.
  A true chaste love hath entered in my breast.
  I love you dearly as I love my life.   50
  I love you as a husband loves a wife.
  That and no other love my love requires.
  Thy wealth, I know, is little; my desires
  Thirst not for gold. Sweet beauteous Jane, what's mine
  Shall, if thou make myself thine, all be thine.   55
  Say, judge, what is thy sentence? Life, or death?
  Mercy or cruelty lies in thy breath.

JANE Good sir, I do believe you love me well,
  For 'tis a silly conquest, silly pride,

For one like you, I mean a gentleman,   60
To boast that by his love tricks he hath brought
Such and such women to his amorous lure.
I think you do not so; yet many do,
And make it even a very trade to woo.
I could be coy, as many women be,   65
Feed you with sunshine smiles and wanton looks.
But I detest witchcraft. Say that I
Do constantly believe you constant have—
HAM. Why dost thou not believe me?
JANE   I believe you.

But yet, good sir, because I will not grieve you   70
With hopes to taste fruit which will never fall,
In simple truth this is the sum of all:
My husband lives, at least I hope he lives.
Pressed he was to these bitter wars in France:
Bitter they are to me by wanting him.   75
I have but one heart and that heart's his due.
How can I then bestow the same on you?
Whilst he lives, his I live, be it ne'er so poor,
And rather be his wife than a king's whore.
HAM. Chaste and dear woman, I will not abuse thee,   80
Although it cost my life if thou refuse me.
Thy husband pressed for France? What was his name?
JANE Rafe Damport.
HAM.   Damport. Here's a letter sent
From France to me from a dear friend of mine,
A gentleman of place. Here he doth write   85
Their names that have been slain in every fight.
JANE I hope death's scroll contains not my love's name.
HAM. Cannot you read?
JANE   I can.
HAM.   Peruse the same.

73

    To my remembrance such a name I read
    Amongst the rest. See here.
JANE   Ay me! He's dead—   90
    He's dead. If this be true, my dear heart's slain.
HAM. Have patience, dear love.
JANE   Hence, hence.
HAM.   Nay, sweet Jane,
    Make not poor sorrow proud with these rich tears.
    I mourn thy husband's death because thou mourn'st.
JANE   That bill is forged. 'Tis sign'd by forgery.   95
HAM. I'll bring thee letters sent besides to many
    Carrying the like report. Jane, 'tis too true.
    Come, weep not: mourning, though it rise from love,
    Helps not the mournèd, yet hurts them that mourn.
JANE   For God's sake, leave me.
HAM.   Whither dost thou turn?   100
    Forget the dead, love them that are alive.
    His love is faded; try how mine will thrive.
JANE   'Tis now no time for me to think on love.
HAM. 'Tis now best time for you to think on love,
    Because your love lives not.
JANE   Though he be dead,   105
    My love to him shall not be buried.
    For God's sake, leave me to myself alone.
HAM. 'Twould kill my soul to leave thee drowned in moan.
    Answer me to my suit and I am gone.
    Say to me yea or no.
JANE   No.
HAM.   Then farewell.   110
    One farewell will not serve—I come again.
    Come, dry these wet cheeks. Tell me, faith, sweet Jane,
    Yea or no, once more.
JANE   Once more I say no.

Once more be gone, I pray, else will I go.

HAM. Nay then, I will grow rude. By this white hand,   115
  Unless you change that cold 'No', here I'll stand,
  Till by your hard heart—

JANE   Nay, for God's sake, peace!
  My sorrows by your presence more increase.
  Not that you thus are present, but all grief
  Desires to be alone. Therefore in brief   120
  Thus much I say, and saying bid adieu:
  If ever I wed man, it shall be you.

HAM. Oh, blessed voice! Dear Jane, I'll urge no more.
  Thy breath hath made me rich.

JANE   Death makes me poor.

                             Exeunt.

ACT IV

SCENE I

London: Eyre's shop (now Hodge's). Enter HODGE *at his shop board,* RAFE, FIRKE, LACY (*as Hans*), *and a boy at work.*

ALL (*singing*) Hey down, a-down, a-down, down derry.

HODGE Well said, my hearts. Ply your work today. We loitered yesterday. To it pell mell that we may live to be Lord Mayors, or Aldermen at least.

FIRKE Hey down, a-down derry.   5

HODGE Well said, i'faith. How say'st thou, Hans? Doth not Firke tickle it?

LACY Yaw, mester.

FIRKE Not so neither. My organ pipe squeaks this morning for want of liquoring. Hey down a-down derry!   10

LACY Forware, Firke, tow best un jolly yongster. Hort I, mester, ic bid yo cut me un pair vampies vor mester Jeffres' bootes.

HODGE Thou shalt, Hans.

FIRKE Master.   15

HODGE How now, boy?

FIRKE Pray, now you are in the cutting vein, cut me out a pair of counterfeits, or else my work will not pass current. Hey down a-down.

HODGE Tell me, sirs, are my cousin Mistress Priscilla's shoes 20 done?

FIRKE Your cousin? No, master: one of your aunts, hang her. Let them alone!

RAFE I am in hand with them. She gave charge that none but I should do them for her.   25

FIRKE Thou do for her? Then 'twill be a lame doing, and

76

that she loves not. Rafe, thou mightst have sent her to me.
In faith, I would have yerked and firked your Priscilla.
Hey down a-down derry. This gear will not hold.

HODGE How say you, Firke? Were we not merry at Old 30
Ford?

FIRKE How, merry? Why, our buttocks went jiggy-joggy
like a quagmire. Well, Sir Roger Oatmeal, if I thought
all meal of that nature, I would eat nothing but bagpuddings.   35

RAFE Of all good fortunes, my fellow Hans had the best.

FIRKE 'Tis true, because Mistress Rose drank to him.

HODGE Well, well, work apace. They say seven of the
Aldermen be dead, or very sick.

FIRKE I care not. I'll be none.   40

RAFE No, nor I. But then my master Eyre will come quickly
to be Lord Mayor.   *Enter* SYBIL.

FIRKE Whoop! Yonder comes Sybil.

HODGE Sybil, welcome, i'faith. And how dost thou, mad
wench?   45

FIRKE Syb whore, welcome to London.

SYBIL Godamercy, sweet Firke. Good lord, Hodge! What
a delicious shop you have got. You tickle it, i'faith.

RAFE Godamercy, Sybil, for our good cheer at Old Ford.

SYBIL That you shall have, Rafe.   50

FIRKE Nay, by the mass, we had tickling cheer, Sybil. And
how the plague dost thou and Mistress Rose? And my
Lord Mayor? (I put the women in first).

SYBIL Well, Godamercy. But God's me, I forget myself!
Where's Hans, the Fleming?   55

FIRKE Hark, butter-box, now you must yelp out some
spreken.

LACY Vat begaie you? Vat vod you, frister?

77

SYBIL   Marry, you must come to my young mistress, to pull
on her shoes you made last.   60

LACY   Vare ben your edle fro? Vare ben your mistress?

SYBIL   Marry, here at our London house in Cornwall.

FIRKE   Will nobody serve her turn but Hans?

SYBIL   No, sir. Come, Hans, I stand upon needles.

HODGE   Why then, Sybil, take heed of pricking.   65

SYBIL   For that let me alone: I have a trick in my budget.
Come, Hans.

LACY   Yaw, yaw, ic sall meet yo gane.

HODGE   Go, Hans, make haste again. Come, who lacks
work?   *Exeunt Lacy and Sybil.* 70

FIRKE   I, master, for I lack my breakfast. 'Tis munching
time, and past.

HODGE   Is't so? Why then, leave work, Rafe. To breakfast.
Boy, look to the tools. Come, Rafe, come Firke.   *Exeunt.*

SCENE 2

The same. Enter a Serving Man.

SERV.   Let me see, now: the sign of the Last in Tower Street.
Mass, yonder's the house. (*shouts*) What, ho! Who's
within?

Enter RAFE.

RAFE   Who calls there? What want you, sir?

SERV.   Marry, I would have a pair of shoes made for a   5
gentlewoman against tomorrow morning. What, can you
do them?

RAFE   Yes, sir, you shall have them. But what length's her
foot?

SERV.   Why, you must make them in all parts like this shoe. 10
But at any hand, fail not to do them, for the gentlewoman
is to be married very early in the morning.

RAFE   How? By *this* shoe must it be made? By this? Are
you sure, sir, by this?

SERV.   How, 'by this am I sure, by this'? Art thou in thy 15
wits? I tell thee I must have a pair of shoes, dost thou
mark me? A pair of shoes, two shoes, made by this very
shoe, this same shoe, against tomorrow morning by four
o'clock. Dost understand me? Canst thou do't?

RAFE   Yes, sir, yes, ay, ay, I can do't. By this shoe, you say. 20
I should know this shoe. Yes, sir, yes, by this shoe. I can
do't...four o'clock...well. Whither shall I bring them?

SERV.   To the sign of the Golden Ball in Watling Street.
Enquire for one Master Hammon, a gentleman, my master.

RAFE   Yea, sir. By this shoe, you say?   25

SERV.   I say Master Hammon at the Golden Ball. He's the
bridegroom and those shoes are for his bride.

RAFE   They shall be done, by this shoe. Well, well...Master
Hammon at the Golden Shoe...I would say the Golden
Ball. Very well, very well. But I pray you, sir, where must 30
Master Hammon be married?

SERV.   At Saint Faith's Church under Paul's, but what's that
to thee? Prithee, dispatch those shoes, and so farewell.

Exit.

RAFE   By this shoe, said he. How am I amazed
At this strange accident! Upon my life,   35
This was the very shoe I gave my wife
When I was pressed for France, since when, alas,
I never could hear of her. It is the same,
And Hammon's bride no other but my Jane.

Enter FIRKE.

FIRKE   'Snails, Rafe! Thou hast lost thy part of three pots a 40
countryman of mine gave me to breakfast.

RAFE   I care not—I have found a better thing.

79

FIRKE A thing? Away! Is it a man's thing or a woman's thing?

RAFE Firke, dost thou know this shoe?   45

FIRKE No, by my troth, neither doth that know me. I have no acquaintance with it, 'tis a mere stranger to me.

RAFE Why, then, I do: this shoe, I durst be sworn,
Once covered the instep of my Jane.
This is her size, her breadth; thus trod my love;   50
These true love knots I pricked. I hold my life,
By this old shoe I shall find out my wife!

FIRKE Ha ha! Old shoe, that wert new—how a murrain came this ague fit of foolishness upon thee?

RAFE Thus, Firke: even now here came a serving man.   55
By this shoe would he have a new pair made
Against tomorrow morning for his mistress
That's to be married to a gentleman.
And why may not this be my sweet Jane?

FIRKE And why may'st not thou be my sweet ass? Ha ha! 60

RAFE Well, laugh and spare not. But the truth is this:
Against tomorrow morning I'll provide
A lusty crew of honest shoemakers
To watch the going of the bride to church.
If she prove Jane, I'll take her in despite   65
From Hammon and the devil, were he by.
If it be not my Jane, what remedy?
Hereof am I sure—I shall live till I die,
Although I never with a woman lie.   *Exit.*

FIRKE Thou lie with a woman to build nothing but Cripple- 70
gates! Well, God sends fools fortune, and it may be he may
light upon his matrimony by such a device. For wedding
and hanging goes by destiny.   *Exit.*

SCENE 3

The Lord Mayor's House in London. Enter LACY
(*as Hans*) *and* ROSE *arm-in-arm.*

LACY How happy am I by embracing thee!
Oh, I did fear such cross mishaps did reign
That I should never see my Rose again.
ROSE Sweet Lacy, since fair opportunity
Offers herself to further our escape,   5
Let not too over-fond esteem of me
Hinder that happy hour: invent the means,
And Rose will follow thee through all the world.
LACY Oh, how I surfeit with excess of joy,
Made happy by thy rich perfection.   10
But since thou pay'st sweet interest to my hopes,
Redoubling love on love, let me once more,
Like to a bold-fac'd debtor, crave of thee
This night to steal abroad, and at Eyre's house,
Who now by death of certain Aldermen   15
Is Mayor of London, and my master once,
Meet thou thy Lacy. Where, in spite of change,
Your father's anger and mine uncle's hate,
Our happy nuptials will we consummate.

Enter SYBIL.

SYBIL Oh God, what will you do, mistress? Shift for your- 20
self—your father is at hand! He's coming, he's coming!
Master Lacy, hide yourself in my mistress. For God's sake,
shift for yourselves!
LACY Your father come, sweet Rose. What shall I do?
Where shall I hide me? How shall I escape?   25
ROSE A man, and want wit in extremity?
Come, come, be Hans still; play the shoemaker;
Pull on my shoe.

6   81   SD

Act IV, Scene 3

Enter the former LORD MAYOR.

LACY Mass, that's well remembered.

SYBIL Here comes your father.   30

LACY Forware, metress, tis un good skow. It sal vel dute,
or ye sall neit betallen.

ROSE Oh God, it pincheth me! (*to Lacy*) What will you do?

LACY (*aside to Rose*) Your father's presence pincheth, not
the shoe.

L.MA. Well done, fit my daughter well and she shall please 35
thee well.

LACY Yaw, yaw, ick weit dat well. Forware, tis un good
skoo. Tis gemait van neit's leather, se ever mine here.

Enter a Prentice.

L.MA. I do believe it. What's the news with you?

PRENT. Please you, the Earl of Lincoln at the gate   40
Is newly lighted and would speak with you.

L.MA. The Earl of Lincoln come to speak with me?
Well, well, I know his errand. Daughter Rose,
Send hence your shoemaker. Dispatch, have done.
Syb, make things handsome. Sir boy, follow me.   45

Exeunt Lord Mayor, Sybil and Prentice.

LACY Mine uncle come! Oh, what may this portend?
Sweet Rose, this of our love threatens an end.

ROSE Be not dismayed at this. What e're befall,
Rose is thine own. To witness I speak truth,
Where thou appoints the place, I'll meet with thee.   50
I will not fix a day to follow thee,
But presently steal hence—do not reply:
Love which gave strength to bear my father's hate
Shall now add wings to further our escape.   *Exeunt.*

SCENE 4

The same. Enter (former) LORD MAYOR *and* LINCOLN.

L.MA. Believe me, on my credit, I speak truth.
　Since first your nephew Lacy went to France
　I have not seen him. It seemed strange to me,
　When Dodger told me that he stayed behind,
　Neglecting the high charge the King imposed.　　　5
LINC. Trust me, Sir Roger Oatley, I did think
　Your counsel had given head to this attempt,
　Drawn to it by the love he bears your child.
　Here I did hope to find him, in your house.
　But now I see mine error and confess　　　10
　My judgment wronged you by conceiving so.
L.MA. Lodge in my house, say you? Trust me, my Lord,
　I love your nephew Lacy too too dearly
　So much to wrong his honour; and he hath done so
　That first gave him advice to stay from France.　　　15
　To witness I speak truth, I let you know
　How careful I have been to keep my daughter
　Free from all conference or speech of him.
　Not that I scorn your nephew, but in love
　I bear your honour, lest your noble blood　　　20
　Should by my mean worth be dishonourèd.
LINC. (*aside*) How far the churl's tongue wanders from his
　　heart.
　Well, well, Sir Roger Oatley, I believe you,
　With more than many thanks for the kind love
　So much you seem to bear me. But, my Lord,　　　25
　Let me request your help to seek my nephew,
　Whom, if I find, I'll straight embark for France.
　So shall your Rose be free, my thoughts at rest,
　And much care die which now lies in my breast.

83　　　　　　6-2

Enter SYBIL.

SYBIL   Oh Lord, help, for God's sake! My mistress! Oh, 30
my young mistress!

L.MA.   Where is thy mistress? What's become of her?

SYBIL   She's gone! She's fled!

L.MA.   Gone! Whither is she fled?

SYBIL   I know not, forsooth.... She's fled out of doors with 35
Hans, the shoemaker. I saw them scud, scud, scud, apace,
apace.

L.MA.   Which way? What, John! Where be my men?
Which way?

SYBIL   I know not, and it please your Worship.   40

L.MA.   Fled with a shoemaker! Can this be true?

SYBIL   Oh Lord, sir, as true as God's in heaven.

LINC.   (*aside*) Her love turned shoemaker? I am glad of
this.

L.MA.   A Fleming butter-box, a shoemaker?   45
Will she forget her birth? Requite my care
With such ingratitude? Scorned she young Hammon
To love a honnikin, a needy knave?
Well, let her fly, I'll not fly after her.
Let her starve if she will, she's none of mine.   50

LINC.   Be not so cruel, sir.

Enter FIRKE *with shoes.*

SYBIL   (*aside*) I am glad she's scaped.

L.MA.   I'll not account of her as of my child.
Was there no better object for her eyes
But a foul drunken lubber, swillbelly,   55
A shoemaker? That's brave!

FIRKE   Yea, forsooth, 'tis a very brave shoe, and as fit as a
pudding.

84

L.MA. How now? What knave is this? From whence comest
thou?   60

FIRKE No knave, sir. I am Firke, the shoemaker; lusty
Roger's chief lusty journeyman, and I come hither to take
up the pretty leg of sweet Mistress Rose...and thus hoping
your Worship is in as good health as I was at the making
hereof, I bid you farewell, Yours, Firke.   65

L.MA. Stay, stay, sir knave.

LINC. Come hither, shoemaker.

FIRKE 'Tis happy the knave is put before the shoemaker, or
else I would not have vouchsafed to come back to you.
I am moved, for I stir.   70

L.MA. My Lord, this villain calls us knaves by craft.

FIRKE Then 'tis the Gentle Craft, and to call one knave
gently is no harm. Sit your Worship merry. (*aside to Sybil*)
Syb, your young mistress...I'll so bob them now my
Master Eyre is Lord Mayor of London.   75

L.MA. Tell me, sirrah, whose man are you?

FIRKE I am glad to see your worship so merry...I have no
maw to this gear, no stomach as yet to a red petticoat.
                                                                (*Pointing to Sybil*)

LINC. He means not, sir, to woo you to his maid,
But only doth demand whose man you are.   80

FIRKE I sing now to the tune of Rogero; Roger, my fellow,
is now my master.

LINC. Sirrah, know'st thou one Hans, a shoemaker?

FIRKE Hans, shoemaker? Oh yes...stay...yes, I have him!
I tell you what (I speak it in secret): Mistress Rose and he 85
are by this time.... No, not so; but shortly are to come
over one another with 'Can you dance the shaking of the
sheets?' It is that Hans. (*aside*) I'll so gull these diggers!

L.MA. Know'st thou then where he is?

FIRKE Yea, forsooth. Yea, marry.   90

LINC. Canst thou? In sadness?

FIRKE No, forsooth. No, marry.

L.MA. Tell me, good honest fellow, where he is,
And thou shalt see what I'll bestow of thee.

FIRKE Honest fellow? No, sir, not so, sir. My profession 95
is the Gentle Craft. I care not for seeing: I love feeling.
Let me feel it here. Aurium tenus, ten pieces of gold:
genuum tenus, ten pieces of silver. And then Firke is your
man on a new pair of stretchers.

L.MA. Here is an angel, part of thy reward, 100
Which I will give thee. Tell me where he is.

FIRKE No, point. Shall I betray my brother? No. Shall I
prove Judas to Hans? No. Shall I cry treason to my corporation? No. I shall be firked and yerked then. But give
me your angel; your angel shall tell you. 105

LINC. Do so, good fellow. 'Tis no hurt to thee.

FIRKE Send simpering Syb away.

L.MA. Huswife, get you in.   *Exit Sybil.*

FIRKE Pitchers have ears, and maids have wide mouths. But
for Hans-prans, upon my word, tomorrow morning he 110
and young Mistress Rose go to this gear. They shall be
married together, by this rush, or else turn Firke to a
firkin of butter to tan leather withal.

L.MA. But art thou sure of this?

FIRKE Am I sure that Paul's steeple is a handful higher than 115
London Stone? Or that the Pissing Conduit leaks nothing
but pure Mother Bunch? Am I sure that I am lusty Firke?
God's nails! Do you think I am so base to gull you?

LINC. Where are they married? Dost thou know the church?

FIRKE I never go to church, but I know the name of it. It is 120
a swearing church. Stay a while...'tis...Ay! By the mass
...no, no, 'tis...ay, by my faith, that that...'tis...Ay!
By my Faith's Church under Paul's Cross! There they shall

be knit like a pair of stockings in matrimony. There they'll
be incony.   125

LINC. Upon my life, my nephew Lacy walks
  In the disguise of this Dutch shoemaker.

FIRKE Yes, forsooth.

LINC. Doth he not, honest fellow?

FIRKE No, forsooth. I think Hans is nobody but Hans. No   130
  spirit.

L.MA. My mind misgives me now, 'tis so indeed.

LINC. My cousin speaks the language, knows the trade.

L.MA. Let me request your company, my Lord.
  Your honourable presence may, no doubt,   135
  Refrain their headstrong rashness, when myself
  Going alone perchance may be o'erborne.
  Shall I request this favour?

LINC.   This, or what else.

FIRKE Then you must rise betimes, for they mean to fall to
  their hey pass and repass, pindy pandy, which hand will 140
  you have, very early.

L.MA. My care shall every way equal their haste.
  This night accept your lodging in my house.
  The earlier shall we stir, and at Saint Faith's
  Prevent this giddy, hair-brained nuptial.   145
  This traffic of hot love shall yield cold gains.
  They ban our loves, and we'll forbid their banns.   *Exit.*

LINC. At Saint Faith's Church, thou say'st?

FIRKE Yes, by their troth.

LINC. Be secret on thy life.   *Exit.* 150

FIRKE Yes, when I kiss your wife! Ha, ha! Here's no craft
  in the Gentle Craft. I came hither of purpose with shoes to
  Sir Roger's worship, whilst Rose, his daughter, be coney-
  catched by Hans. Soft now! These two gulls will be at
  Saint Faith's Church tomorrow morning, to take Master   155

Bridegroom and Mistress Bride napping. And they in the meantime shall chop up the matter at the Savoy. But the best sport is: Sir Roger Oatley will find my fellow, lame Rafe's wife going to marry a gentleman, and then he'll stop her instead of his daughter. Oh brave! There will be 160 fine tickling sport. Soft now, what have I to do? Oh, I know now. A mess of shoemakers meet at the Woolsack in Ivy Lane to cozen my gentleman of lame Rafe's wife, that's true.

Alack, alack!   165
Girls, hold out tack!
For now smocks for this jumbling
Shall go to wrack.   *Exit.*

ACT V

Eyre's house. Enter EYRE, *his* WIFE, LACY
(*as Hans*) *and* ROSE.

EYRE   This is the morning, then. Stay my bully, my honest
Hans: is it not?

LACY   This is the morning that must make us two
Happy or miserable. Therefore, if you—

EYRE   Away with these if's and an's, Hans, and these et   5
ceteras. By mine honour, Rowland Lacy, none but the
King shall wrong thee. Come, fear nothing: am not I Sim
Eyre? Is not Sim Eyre Lord Mayor of London? Fear
nothing, Rose. Let them all say what they can. Dainty,
come thou to me. Laughest thou?   10

WIFE   Good my Lord, stand her friend in what thing you
may.

EYRE   Why, my sweet Lady Madgy, think you Simon Eyre
can forget his fine Dutch journeyman? No, vah. Fie, I
scorn it! It shall never be cast in my teeth that I was   15
unthankful. Lady Madgy, thou hadst never covered thy
Saracen's head with this French flap, nor loaden thy bum
with this farthingale...'tis trash, trumpery, vanity...
Simon Eyre had never walked in a red petticoat, nor worn
a chain of gold, but for my fine journeyman's Portuguese.   20
And shall I leave him? No! Prince am I none, yet bear a
princely mind.

LACY   My Lord, 'tis time for us to part from hence.

EYRE   Lady Madgy, Lady Madgy, take two or three of my
piecrust eaters, my buff-jerkin varlets that do walk in black   25
gowns at Simon Eyre's heels, take them, good Lady Madgy.

Trip and go, my brown Queen of Perriwigs, with my
delicate Rose and my jolly Rowland, to the Savoy. See
them linked, countenance the marriage, and, when it is
done, cling, cling together, you Hamborough turtle doves. 30
I'll bear you out. Come to Simon Eyre, come dwell with
me, Hans. Thou shalt eat minced pies and marchpane,
Rose. Away, cricket! Trip and go! My Lady Madgy, to
the Savoy! Hans, wed and to bed; kiss and away. Go,
vanish!   35

WIFE   Farewell, my Lord.

ROSE   Make haste, sweet love.

WIFE   She'd fain the deed were done.

LACY   Come, my sweet Rose. Faster than deer we'll run.

They go out.

EYRE   Go, vanish, vanish! Avaunt, I say! By the Lord of 40
Ludgate, it's a mad life to be a Lord Mayor. It's a stirring
life, a fine life, a velvet life, a careful life. Well, Simon Eyre,
set a good face on it, in the honour of Saint Hugh. Soft!
The king this day comes to dine with me, to see my new
buildings. His Majesty is welcome: he shall have good 45
cheer, delicate cheer, princely cheer. This day my fellow
prentices of London come to dine with me too: they shall
have fine cheer, gentlemanlike cheer. I promised the mad
Cappidosians, when we all served at the Conduit together,
that if ever I came to be Mayor of London I would feast 50
them all, and I'll do't. I'll do't, by the life of Pharoah!
By this beard, Sim Eyre will be no flincher! Besides, I have
procured that upon every Shrove Tuesday, at the sound of
the pancake bell, my fine dapper Assyrian lads shall clap up
their shop windows, and away. This is the day! And this 55
day they shall do't, they shall do't...!
Boys, that day are you free. Let masters care,
And prentices shall pray for Simon Eyre.   *Exit.*

SCENE 2

A street near St Faith's Church. Enter HODGE, FIRKE, RAFE
and five or six shoemakers, all with cudgels or such weapons.

HODGE   Come, Rafe; stand to it, Firke. My masters, as we
are the brave bloods of the shoemakers, heirs apparent to
Saint Hugh, and perpetual benefactors to all good fellows,
thou shalt have no wrong. Were Hammon a king of spades,
he should not delve in thy close without thy sufferance.   5
But tell me, Rafe, art thou sure 'tis thy wife?

RAFE   Am I sure this is Firke? This morning, when I stroked
on her shoes, I looked upon her, and she upon me, and
sighed, asked me if ever I knew one Rafe. 'Yes,' said I.
'For his sake', said she, tears standing in her eyes, 'and for   10
thou art somewhat like him, spend this piece of gold.' I
took it. My lame leg and my travel beyond sea made me
unknown. All is one for that: I know she's mine.

FIRKE   Did she give thee this gold? Oh glorious glittering
gold! She's thine own, 'tis thy wife and she loves thee; for   15
I'll stand to't, there's no woman will give gold to any man,
but she thinks better of him than she thinks of them she
gives silver to. And for Hammon, neither Hammon nor
hangman shall wrong thee in London. Is not our old
master Eyre, Lord Mayor? Speak, my hearts.   20

ALL   Yes, and Hammon shall know it to his cost.

Enter HAMMON, *his man,* JANE, *and others.*

HODGE   Peace, my bullies. Yonder they come.

RAFE   Stand to't, my hearts. Firke, let me speak first.

HODGE   No, Rafe, let me. (*comes forward*) Hammon, whither
away so early?   25

HAM.   Unmannerly, rude slave, what's that to thee?

FIRKE   To *him*, sir? Yes, sir, and to me, and others. Good-

morrow, Jane, how dost thou? Good Lord, how the world
is changed with you, God be thanked.

HAM. Villains, hands off! How dare you touch my love? 30

ALL THE SHOEMAKERS Villains? Down with them! Cry
clubs for prentices!

HODGE Hold, my hearts. Touch her, Hammon? Yea, and
more than that, we'll carry her away with us. My masters
and gentlemen, never draw your bird-spits. Shoemakers 35
are steel to the back, men every inch of them, all spirit.

ALL OF HAMMON'S SIDE Well, and what of all this?

HODGE I'll show you. Jane, dost thou know this man? 'Tis
Rafe, I can tell thee. Nay, 'tis he in faith, though he be
lamed by the wars. Yet look not strange, but run to him. 40
Fold him about the neck and kiss him.

JANE Lives then my husband? Oh, God, let me go,
Let me embrace my Rafe!

HAM.   What means my Jane?

JANE Nay, what meant you to tell me he was slain?

HAM. Pardon me, dear love, for being misled.   45
(*to Rafe*) 'Twas rumoured here in London thou wert
dead.

FIRKE Thou seest he lives. Lass, go pack home with him.
Now, Master Hammon, where's your mistress, your wife?

SERVANT 'Swounds, master! Fight for her! Will you thus
lose her?   50

SHOEMAKERS Down with that creature! Clubs! Down
with him!

HODGE Hold, hold!

HAM. Hold, fool! Sirs, he shall do no wrong.
(*to Jane*) Will my Jane leave me thus, and break her faith? 55

FIRKE Yea, sir. She must, sir, she shall, sir. What then?
Mend it.

HODGE Hark, fellow Rafe, follow my counsel. Set the

wench in the midst and let her choose her man and let her
be his woman.   60

JANE   Whom should I choose? Whom should my thoughts
affect
But him whom heaven hath made to be my love?
Thou art my husband, and these humble weeds
Makes thee more beautiful than all his wealth.
Therefore I will but put off his attire,   65
Returning it into the owner's hand,
And after ever be thy constant wife.

HODGE   Not a rag, Jane. The law's on our side. He that
sows in another man's ground forfeits his harvest. Get thee
home, Rafe. Follow him, Jane. He shall not have so much 70
as a busk-point from thee.

FIRKE   Stand to that, Rafe. The appurtenances are thine own.
Hammon, look not at her.

SERVANT   Oh, 'swounds, no.

FIRKE   Bluecoat, be quiet. We'll give you a new livery else. 75
We'll make Shrove Tuesday Saint George's Day for you.
Look not, Hammon, leer not. I'll firke you—for thy head
now, one glance, one sheep's eye, anything at her. Touch
not a rag, lest I and my brethren beat you to clouts.

SERVANT   Come, Master Hammon, there's no striving here. 80

HAM.   Good fellows, hear me speak. And honest Rafe,
Whom I have injured most by loving Jane,
Mark what I offer thee. Here in fair gold
Is twenty pound. I'll give it thee for thy Jane.
If this content thee not, thou shalt have more.   85

HODGE   Sell not thy wife, Rafe. Make her not a whore.

HAM.   Say, wilt thou freely cease thy claim in her,
And let her be my wife?

SHOEMAKERS   No, do not, Rafe!

RAFE   Sirrah Hammon. Hammon, dost thou think a shoe- 90

93

maker is so base to be a bawd to his own wife for commodity? Take thy gold. Choke with it! Were I not lame,
I would make thee eat thy words.

FIRKE   A shoemaker sell his flesh and blood! Oh indignity!

HODGE   Sirrah, take up your pelf and be packing.   95

HAM.   I will not touch one penny. But in lieu
Of that great wrong I offerèd thy Jane,
To Jane and thee I give that twenty pound.
Since I have failed of her, during my life
I vow no woman else shall be my wife.   100
Farewell, good fellows of the Gentle Trade.
Your morning's mirth my mourning day hath made.
                          Exeunt Hammon and Servants.

FIRKE   (*to servant going out*) Touch the gold, creature, if you
dare. Y'are best be trudging. Here, Jane, take thou it. Now
let's home, my hearts.   105

HODGE   Stay, who comes here? Jane, on again with thy
mask!

Enter LINCOLN, (*former*) LORD MAYOR, *and servants.*

LINC.   Yonder's the lying varlet mocked us so.

L.MA.   Come hither, sirrah.

FIRKE   I, sir? I am sirrah? You mean me, do you not?   110

LINC.   Where is my nephew married?

FIRKE   Is he married? God give him joy, I am glad of it.
They have a fair day and the sign is in a good planet—Mars
in Venus.

L.MA.   Villain! Thou told'st me that my daughter Rose,   115
This morning should be married at Saint Faith's.
We have watched there these three hours at the least,
Yet see we no such thing.

FIRKE   Truly, I am sorry for't. A bride's a pretty thing.

HODGE   Come, to the purpose. Yonder's the bride and bride-   120

groom you look for, I hope. Though you be lords, you are
not to bar, by your authority, men from women, are
you?

L.MA. See, see! My daughter's masked.

LINC.   True, and my nephew,
To hide his guilt, counterfeits him lame.   125

FIRKE Yea truly, God help the poor couple! They are lame
and blind.

L.MA. I'll ease her blindness.

LINC.   I'll his lameness cure.

FIRKE (*to the Shoemakers*) Lie down, sirs, and laugh. My
fellow Rafe is taken for Rowland Lacy, and Jane for 130
Mistress Damask Rose. This is all my knavery.

L.MA. (*to Jane*) What, have I found you, minion?

LINC. (*to Rafe*)   Oh, base wretch.
Nay, hide thy face. The horror of thy guilt
Can hardly be washed off. Where are thy powers?
What battles have you made? Oh yes, I see   135
Thou fought'st with shame, and shame hath conquer'd thee.
This lameness will not serve.

L.MA.   Unmask yourself.

LINC. Lead home your daughter.

L.MA.   Take your nephew hence.

RAFE Hence? 'Swounds, what mean you? Are you mad?
I hope you cannot enforce my wife from me. Where's 140
Hammon?

L.MA. Your wife?

LINC. What Hammon?

RAFE Yea, my wife. And therefore, the proudest of you that
lays hands on her first, I'll lay my crutch cross his pate. 145

FIRKE To him, lame Rafe! Here's brave sport!

RAFE Rose call you her? Why, her name is Jane. Look here
else. Do you know her now?   *Unmasks her.*

LINC. Is this your daughter?

L.MA.   No, nor this your nephew.
My Lord of Lincoln, we are both abused   150
By this base crafty varlet.

FIRKE Yea, forsooth, no varlet. Forsooth, no base. Forsooth,
I am but mean. Not crafty neither, but of the Gentle Craft.

L.MA. Where is my daughter Rose? Where is my child?

LINC. Where is my nephew Lacy married?   155

FIRKE Why, here is good laced mutton, as I promised you.

LINC. Villain! I'll have thee punished for this wrong.

FIRKE Punish the journeyman villain, but not the journeyman shoemaker.
                    Enter DODGER.

DODGER My Lord, I come to bring unwelcome news: 160
Your nephew Lacy and your daughter Rose
Early this morning wedded at the Savoy,
None being present but the Lady Mayoress.
Besides I learnt among the officers
The Lord Mayor vows to stand in their defence   165
Gainst any that shall seek to cross the match.

LINC. Dares Eyre the shoemaker uphold the deed?

FIRKE Yes, sir, shoemakers dare stand in a woman's quarrel,
I warrant you, as deep as another, and deeper too.

DODGER Besides, his Grace today dines with the Mayor,   170
Who on his knees humbly intends to fall
And beg a pardon for your nephew's fault.

LINC. But I'll prevent him. Come, Sir Roger Oatley.
The King will do us justice in this cause.
Howe'er their hands have made them man and wife,   175
I will disjoin the match, or lose my life.
                              Exeunt Lincoln and Oatley.

FIRKE Adieu, Monsieur Dodger, farewell, fools. Ha, ha!
Oh, if they had stayed I would have so lammed them with

flouts. Oh, heart! My codpiece point is ready to fly in pieces every time I think upon Mistress Rose. But let that 180 pass, as my Lady Mayoress says.

HODGE This matter is answered. Come, Rafe, home with thy wife. Come, my fine shoemakers. Let's to our master's, the new Lord Mayor, and there swagger this Shrove Tuesday. I'll promise you wine enough, for Madge keeps 185 the cellar.

ALL Oh rare! Madge is a good wench.

FIRKE And I'll promise you meat enough, for simpering Susan keeps the larder. I'll lead you to victuals, my brave soldiers. Follow your captain. Oh, brave! Hark! Hark! 190

Bell rings.

ALL The pancake bell rings! The pancake bell! Tri-lil, my hearts!

FIRKE Oh brave, oh sweet bell! Oh, delicate pancakes! Open the doors, my hearts, and shut up the windows. Keep in the house, let out the pancakes. Oh rare, my hearts, 195 let's march together for the honour of Saint Hugh to the great new hall in Gracious Street corner, which our master, the new Lord Mayor, hath built.

RAFE Oh, the crew of good fellows that will dine at my Lord Mayor's cost today! 200

HODGE By the Lord, the Lord Mayor is a most brave man. How shall prentices be bound to pray for him, and the honour of the gentlemen shoemakers! Let's feed and be fat with my Lord's bounty.

FIRKE Oh musical bell still! Oh Hodge! Oh my brethren! 205 There's cheer for the heavens. Venison pasties walk up and down piping hot like sergeants. Beef and brewis comes marching in dry fats. Fritters and pancakes comes trowling in in wheelbarrows. Hens and oranges hopping in porters'

baskets, collops and eggs in scuttles, and tarts and custards 210
come quavering in in malt shovels.

Enter more prentices.

ALL   Whoop! Look here, look here!

HODGE   How now, mad lads? Whither away so fast?

I PREN.   Whither? Why, to the great new hall. Know you
not why? The Lord Mayor hath bidden all the prentices in 215
London to breakfast this morning.

ALL   Oh brave shoemaker! Oh brave lord of incomprehen-
sible good fellowship! Whoo! Hark you...the pancake
bell rings.   *They cast up caps.*

FIRKE   Nay, more, my hearts. Every Shrove Tuesday is 220
our year of jubilee. And when the pancake bell rings,
we are as free as my Lord Mayor. We may shut up our
shops and make holiday. I'll have it called Saint Hugh's
Holiday.

ALL   Agreed, agreed! Saint Hugh's Holiday!   225

HODGE   And this shall continue for ever.

ALL   Oh brave! Come, come, my hearts. Away, away!

FIRKE   Oh, eternal credit to us of the Gentle Craft. March
fair, my hearts! Oh rare!   *Exeunt.*

SCENE 3

A street in London. Enter KING
and his train over the stage.

KING   Is our Lord Mayor of London such a gallant?

NOBLEMAN   One of the merriest madcaps in your land.
Your Grace will think when you behold the man,
He's rather a wild ruffian than a Mayor.
Yet thus much I'll ensure your Majesty,   5
In all his actions that concern his state

98

He is as serious, provident and wise,
As full of gravity amongst the grave,
As any Mayor hath been these many years.
KING   I am with child till I behold this huff-cap.   10
But all my doubt is, when we come in presence
His madness will be dashed clean out of countenance.
NOBLEMAN   It may be so, my Liege.
KING   Which to prevent,
Let someone give him notice 'tis our pleasure
That he put on his wonted merriment.   15
Set forward.
ALL   On afore!   *Exeunt.*

SCENE 4

A great hall. Enter EYRE, HODGE, FIRKE, RAFE *and
other shoemakers, all with napkins on their shoulders.*

EYRE   Come, my fine Hodge, my jolly gentlemen shoemakers. Soft, where be these cannibals, these varlets my
officers? Let them all walk and wait upon my brethren;
for my meaning is that none but shoemakers, none but the
livery of my Company, shall in their satin hoods wait upon 5
the trencher of my sovereign.
FIRKE   Oh my Lord, it will be rare!
EYRE   No more, Firke. Come lively. Let your fellow
prentices want no cheer. Let wine be plentiful as beer and
beer as water. Hang these penny-pinching fathers that 10
cram wealth in innocent lamb-skins. Rip, knaves! Avaunt!
Look to my guests.
HODGE   My Lord, we are at our wits' end for room. Those
hundred tables will not feast the fourth part of them.
EYRE   Then cover me those hundred tables again, and again 15
till all my jolly prentices be feasted. Avoid, Hodge. Run,

Rafe. Frisk about, my nimble Firke. Carouse me fathom healths to the honour of the shoemakers. Do they drink lively, Hodge? Do they tickle it, Firke?

FIRKE Tickle it? Some of them have taken their liquor 20 standing so long that they can stand no longer. But for meat, they would eat it an they had it.

EYRE Want they meat? Where's this swag-belly, this greasy kitchen-stuff cook? Call my varlet to me. Want meat! Firke, Hodge, lame Rafe! Run, my tall men, beleaguer the 25 shambles. Beggar all Eastcheap, serve me whole oxen in chargers and let sheep whine upon the table like pigs for want of good fellows to eat them. Want meat! Vanish, Firke. Avaunt, Hodge.

HODGE Your Lordship mistakes my man Firke. He means 30 their bellies want meat, not the boards, for they have drunk so much they can eat nothing.

Enter LACY (*in the attire of Hans*), ROSE *and Eyre's* WIFE.

WIFE Where is my lord?

EYRE How now, Lady Madgy?

WIFE The King's most excellent Majesty is new done. He 35 sends me for thy honour. One of his most worshipful peers bade me tell thou must be merry and so forth. But let that pass.

EYRE Is my Sovereign come? Vanish, my tall shoemakers, my nimble brethren. Look to my guests, the prentices— 40 yet stay a little. How now, Hans? How looks my little Rose?

LACY Let me request you to remember me.
I know your Honour easily may obtain
Free pardon of the King for me and Rose,   45
And reconcile me to my uncle's grace.

EYRE Have done, my good Hans. My honest journeyman,

look cheerly. I'll fall upon both my knees till they be as
hard as horn, but I'll get thy pardon.

WIFE Good my Lord, have a care what you speak to his 50
Grace.

EYRE Away, you Islington whitepot. Hence, you happer-
arse, you barley pudding full of maggots! You broiled
carbonado, avaunt! Avaunt, avoid, Mephostophilus! Shall
Sim Eyre learn to speak of you, Lady Madgy? Vanish, 55
Mother Miniver-Cap, vanish. Go! Trip and go. Meddle
with your partlets and your pishery-pashery, your flues
and your whirligigs. Go! Rub! Out of mine alley! Sim
Eyre knows how to speak to a Pope, to Sultan Soliman, to
Tamburlaine an he were here. And shall I melt, shall 60
I droop before my Sovereign? No. Come, my Lady
Madgy; follow me, Hans. About your business, my frolic
free-booters. Firke, frisk about and about and about, for
the honour of mad Simon Eyre, Lord Mayor of London.

FIRKE Hey for the honour of the shoemakers!   *Exeunt.* 65

SCENE 5

An open yard before the Hall. A long flourish or two. Enter KING,
Nobles, EYRE, *his* WIFE, LACY *(as himself) and* ROSE. *Lacy
and Rose kneel.*

KING Well, Lacy, though the fact was very foul
Of your revolting from our kingly love
And your own duty, yet we pardon you.
Rise both, and Mistress Lacy, thank my Lord Mayor
For your young bridegroom here.   5

EYRE So, my dear Liege, Sim Eyre and my brethren, the
gentlemen shoemakers, shall set your sweet Majesty's image
cheek by jowl by Saint Hugh for this honour you have done
poor Simon Eyre. I beseech your Grace pardon my rude

behaviour. I am a handicrafts man, yet my heart is without 10
craft. I would be sorry at my soul that my boldness should
offend my King.

KING Nay, I pray thee, good Lord Mayor, be even as merry
As if thou wert among thy shoemakers.
It does me good to see thee in this humour. 15

EYRE Say'st thou so, my sweet Dioclesian? Then hump!
Prince am I none, yet am I princely born. By the Lord of
Ludgate, my Liege, I'll be as merry as a pie.

KING Tell me, in faith, mad Eyre, how old thou art.

EYRE My Liege, a very boy, a stripling, a yonker. You see 20
not a white hair on my head, not a gray in this beard.
Every hair, I assure thy Majesty, that sticks in this beard,
Sim Eyre values at the King of Babylon's ransom. Tamar
Cham's beard was a rubbing brush to't. Yet I'll shave
it off and stuff tennis balls with it to please my bully 25
King.

KING But all this while I do not know your age.

EYRE My Liege, I am six and fifty year old. Yet can I cry
hump with a sound heart for the honour of Saint Hugh.
Mark this old wench, my King: I danced the shaking of 30
the sheets with her six and thirty years ago, and yet I hope
to get two or three young Lord Mayors ere I die. I am
lusty still, Sim Eyre still. Care and cold lodging bring
white hairs. My sweet Majesty, let care vanish! Cast it
upon thy nobles, it will make thee always young like 35
Apollo. And cry hump: Prince am I none, yet am I
princely born.

KING Ha ha! Say, Cornwall, didst thou ever see his like?

NOBLEMAN Not I, my Lord.

Enter LINCOLN *and (former)* LORD MAYOR.

KING Lincoln, what news with you?

LINC. My gracious Lord, have care unto yourself,   40
   For there are traitors here.
ALL   Traitors? Where? Who?
EYRE   Traitors in my house? God forbid! Where be my
   officers? I'll spend my soul ere my King feel harm.
KING   Where is the traitor, Lincoln?
LINC. (*pointing to Lacy*) Here he stands.   45
KING   Cornwall, lay hold on Lacy. Lincoln, speak.
   What canst thou lay unto thy nephew's charge?
LINC. This, my dear Liege. Your Grace, to do me honour,
   Heaped on the head of this degenerous boy
   Desertless favours. You made choice of him   50
   To be commander over powers in France.
   But he—
KING   Good Lincoln, prithee pause awhile.
   Even in thine eyes I read what thou wouldst speak.
   I know how Lacy did neglect our love,   55
   Ran himself deeply, in the highest degree,
   Into vile treason.
LINC.   Is he not a traitor?
KING   Lincoln, he was. Now have we pardoned him.
   'Twas not a base want of true valour's fire
   That held him out of France, but love's desire.   60
LINC. I will not bear his shame upon my back.
KING   Nor shalt thou, Lincoln. I forgive you both.
LINC. Then, good my Liege, forbid the boy to wed
   One whose mean birth will much disgrace his bed.
KING   Are they not married?
LINC.   No, my Liege.
ROSE AND LACY   We are!   65
KING   Shall I divorce them? Oh, be it far
   That any hand on earth should dare untie
   The sacred knot knit by God's majesty.

103

I would not for my crown disjoin their hands
That are conjoined in holy nuptial bands.   70
  How say'st thou, Lacy? Wouldst thou lose thy Rose?
LACY Not for all India's wealth, my Sovereign.
KING But Rose, I am sure, her Lacy would forgo.
ROSE If Rose were ask'd that question, she'd say No.
KING You hear them, Lincoln?
LINC.   Yes, my Liege, I do.   75
KING Yet canst thou find i'the heart to part these two?
  Who seeks, besides you, to divorce these lovers?
L.MA. I do, my gracious Lord, I am her father.
KING Sir Roger Oatley, our last Mayor, I think?
NOBLEMAN The same, my Liege.   80
KING   Would you offend love's laws?
  Well, you shall have your wills. You sue to me
  To prohibit the match. Soft: let me see.
  You are both married, Lacy, art thou not?
LACY I am, dread Sovereign.
KING   Then, upon thy life,
  I charge thee not to call this woman wife.   85
L.MA. I thank your Grace.
ROSE   Oh my most gracious Lord!
                        She kneels.
KING Nay, Rose, never woo me. I tell you true,
  Although as yet I am a bachelor
  Yet I believe I shall not marry you.
ROSE Can you divide the body from the soul,   90
  Yet make the body live?
KING   Yea, so profound?
  I cannot, Rose, but you I must divide.
  Fair maid, this bridegroom cannot be your bride.
  Are you pleased, Lincoln? Oatley, are you pleased?
BOTH Yes, my Lord.

KING   Then must *my* heart be eased.   95
  For, credit me, my conscience lives in pain
  Till these whom I divorced be joined again.
  Lacy, give me thy hand. Rose, lend me thine.
  Be what you would be. Kiss now—so, that's fine.
  At night, lovers, to bed. Now let me see,   100
  Which of you all mislikes this harmony?
L.MA. Will you then take from me my child perforce?
KING   Why, tell me, Oatley, shines not Lacy's name
  As bright in the world's eye as the gay beams
  Of any citizen?
LINC.   Yea, but, my gracious Lord,   105
  I do mislike the match far more than he.
  Her blood is too too base.
KING   Lincoln, no more!
  Dost thou not know that love respects no blood,
  Cares not for difference of birth or state?
  The maid is young, well-born, fair, virtuous,   110
  A worthy bride for any gentleman.
  Besides, your nephew for her sake did stoop
  To bare necessity, and, as I hear,
  Forgetting honours and all courtly pleasures,
  To gain her love became a shoemaker.   115
  As for the honour which he lost in France,
  Thus I redeem it. Lacy, kneel thee down.
  Arise, Sir Rowland Lacy. Tell me now,
  Tell me in earnest, Oatley, canst thou chide,
  Seeing thy Rose a Lady and a bride?   120
L.MA. I am content with what your Grace hath done.
LINC. And I, my Liege, since there's no remedy.
KING   Come on then: all shake hands. I'll have you friends.
  Where there is much love, all discord ends.
  What says my mad Lord Mayor to all this love?   125

EYRE   Oh my Liege, this honour you have done to my fine
    journeyman here, Rowland Lacy, and all these favours
    which you have shown to me this day in my poor house,
    will make Simon Eyre live longer by one dozen of warm
    summers more than he should.   130

KING   Nay, my mad Lord Mayor (that shall be thy name),
    If any grace of mine can length thy life,
    One honour more I'll do thee. That new building
    Which at thy cost in Cornhill is erected
    Shall take a name from us. We'll have it called   135
    The Leadenhall, because in digging it
    You found the lead that covereth the same.

EYRE   I thank your Majesty.

WIFE   God bless your Grace.

KING   Lincoln, a word with you.   140

Enter HODGE, FIRKE, RAFE *and more shoemakers.*

EYRE   How now, my mad knaves? Peace! Speak softly!
    Yonder is the King.

KING   With the old troop which there we keep in pay
    We will incorporate a new supply.
    Before one summer more pass o'er my head,   145
    France shall repent England was injurèd.
    What are all those?

EYRE   All shoemakers, my Liege,
    Sometimes my fellows. In their company
    I lived as merry as an emperor.

KING   My mad Lord Mayor, are all these shoemakers?   150

EYRE   All shoemakers, my Liege. All gentlemen of the
    Gentle Craft, true Trojans, courageous cordwainers. They
    all kneel to the shrine of holy Saint Hugh.

ALL THE SHOEMAKERS   God save your Majesty.

KING   Mad Simon, would they any thing with us?   155

EYRE   Mum, mad knaves! Not a word! I'll do't, I warrant
   you. They are all beggars, my Liege, all for themselves. And I for them all on both knees do entreat, that
   for the honour of poor Simon Eyre and the good of his
   brethren, these mad knaves, your Grace would vouch- 160
   safe some privilege to my new Leadenhall: that it may
   be lawful for us to buy and sell leather there two days a
   week.

KING   Mad Sim, I grant your suit. You shall have patent
   To hold two market days in Leadenhall.   165
   Mondays and Fridays, those shall be the times.
   Will this content you?

ALL   Jesus bless your Grace.

EYRE   In the name of these my poor brethren shoemakers,
   I most humbly thank your Grace. But before I rise, seeing 170
   your are in the giving vein and we in the begging, grant
   Sim Eyre one boon more.

KING   What is it, my Lord Mayor?

EYRE   Vouchsafe to taste of a poor banquet that stands
   sweetly waiting for your sweet presence.   175

KING   I shall undo thee, Eyre, only with feasts.
   Already have I been too troublesome,
   Say, have I not?

EYRE   Oh my dear King, Sim Eyre was taken unawares upon
   a day of shroving which I promised long ago to the pren- 180
   tices of London. For an't please your Highness, in time
   past, I bare the water tankard, and my coat sits not a whit
   the worse upon my back. And then upon a morning some
   mad boys (it was a Shrovetide even as 'tis now) gave me
   my breakfast, and I swore then by the stopple of my 185
   tankard, if ever I came to be Lord Mayor of London, I
   would feast all the prentices. This day, my Liege, I did it,
   and the slaves had an hundred tables five times covered.

They are gone home and vanished.
Yet add more honour to the Gentle Trade: 190
Taste of Eyre's banquet, Simon's happy made.
KING   Eyre, I will taste of thy banquet and will say
I have not met more pleasure on a day.
Friends of the Gentle Craft, thanks to you all.
Thanks, my kind Lady Mayoress, for our cheer.   195
Come, lords, awhile let's revel it at home.
When all our sports and banquetings are done,
Wars must right wrongs which Frenchmen have begun.

Exeunt.

FINIS

NOTES

To all good fellows... This is rather like the summary on the jacket of a modern novel, or the key episodes shown in a film trailer: it provides a glimpse of what is offered and so whets the appetite for more. In giving an outline of the plot, it also has a function comparable to that of the dumb-show which often preceded the performance of Elizabethan plays, prefiguring the action which the audience were about to see dramatised in full.

Professors of the Gentle Craft For 'the Gentle Craft' see pp. 135–6. It was a regularly used term for the shoemaker's trade (cf. 'the noble art' as Victorian for boxing), and occurs throughout the play. In the last scene, for example, the King addresses the shoemakers as his 'friends of the Gentle Craft'. It was also the title of the story-book by Thomas Deloney, which was the source of Dekker's play (see p. 135).

2 *conceited* Ingeniously devised (not 'conceit' in our normal sense, but 'conception', a work of fancy and imagination).

3 *my Lord Admiral's Players* One of the leading companies, formed in 1594 by Edward Alleyn, the famous actor. From 1594 to 1603 they acted in about 215 plays, including *Tamburlaine*, *The Spanish Tragedy* and *The Massacre at Paris*.

3–4 *this present Christmas* Cf. the title page of the 1600 edition: 'as it was acted before the Queen's most excellent Majesty on New Years Day at night last'. It would obviously be a suitable Christmas play, with its holiday spirits and the festive banqueting scene at the end.

THE FIRST THREE-MANS SONG

These two songs are reproduced here, coming before the text of the play itself, as they did in the 1600 edition. Neither of them is given a clear place for performance in the play, though the second is directed 'to be sung at the latter end'. Not all editors interpret this very strictly, however: Ernest Rhys, in the Mermaid series, puts it into the Great Hall scene (v, 4) where it certainly goes well enough. The first has no direction at all. Rhys inserts it in III, 3 (or 5 in his version), its cue being Eyre's 'let's be merry whiles we are young' (ll. 25–6) and its completion marked by the Lord Mayor's comment, 'It's well done' (l. 28). Another possibility is after l. 75 of the same scene, where Eyre incites his 'mad lads' to merriment, and 'All go dancing out'. 'Three-mans songs' were probably rounds for three voices, easy to sing and popular enough to be frowned on by professional musicians. A royal charter of 1555 sought to prevent amateurs 'as tayllers, shoemakers, and such others' singing 'songs called Three Mens Songs in the Taverns, Inns, and such other places in this City, and also at weddings'.

8 *her breast against a briar* A reference to the legend that the nightingale pressed her breast against a thorn while singing.

9 *But Oh I spy the cuckoo*

> cf. The cuckoo then on every tree
> Mocks married men, for thus sings he:
> Cuckoo, cuckoo, cuckoo.
> Oh word of fear
> Unpleasing to a married ear.
>
> (*Love's Labour's Lost*, v, 2)

The point is not merely that the cuckoo is an interloper, but that the sound it makes suggests also the word 'cuckold', i.e. a man whose wife has played him false.

THE SECOND THREE-MANS SONG

2 *Saint Hugh* See the note on pp. 135–7. St Hugh was the patron saint of shoemakers and the originator of the phrase 'the Gentle Craft'.

3 *Ill is the weather that bringeth no gain* A shoemaker can be cheerful about bad weather—the worse it is, the better for business.

14 *as often as there be men to drink* The refrain is repeated, with each of the singers drinking, one by one, until all have drunk.

ACT I. Scene 1

5 *my cousin Lacy* Misleading to modern readers, for Lacy is not his cousin at all, but his nephew. From medieval times to the eighteenth century, however, the word was commonly used to

denote a close relationship, 'cos' being a familiar diminutive as in l. 80 of this scene.

6   *Is much affected to* has much affection for.

11   *mean* of low birth.

15   *need not doubt my girl* 'doubt' means fear or distrust.

20   *To travel countries for experience* The young Elizabethan noble-man would often make a 'grand tour' of Europe. 'Travel in the younger sort', says Bacon, 'is a part of education.' He advises travel 'under some tutor or grave servant', and he suggests 'let diaries be brought in use'.

24   *But to see the end* Punctuation and syntax may be confusing. Modern English puts the infinitive to a similar use in such expressions as 'to come to the point'; and whereas we would no doubt begin a new sentence here, Elizabethan punctuation moves the actor forward in a quick delivery of the speech.

27   *embezzled* wasted (cf. 'When they have with riot and prodigality imbezzelled their estates', Burton, *Anatomy of Melancholy*).

33   *consume me* spend.

34   *And make him* Even if you make him.

35   *rioting* lavish spending.

52   *presently* immediately.

53   *powers* troops.

    He would not for a million A roundabout way of saying that he (his Highness) is very anxious that his troops should have arrived in Dieppe within the next four days ('Not for a million pounds would he accept any other idea than...').

58   *lie at Mile End* This, with Finsbury (cf. l. 61) and Tothill Fields l. 61), was one of the famous training grounds. There are references in Shakespeare (*2 Henry IV*, III, 2, 298 for instance); and the burlesque drill scene in *The Knight of the Burning Pestle* also takes place at Mile End. Finsbury was particularly famous for its archery ground.

62   *With frolic spirits* For 'frolic' the *O.E.D.* gives 'joyous, merry, mirthful...sportive, full of merry pranks'.

63   *their imprest* advance payment.

    furniture equipment (i.e. what they are 'furnished' or provided with).

71   *To approve your loves to me? No! Subtlety!* Needs sensible reading. The tone of the question is ironical; the rest means that the Lord Mayor's polite talk is all so much eyewash.

84 *And yet not thee* In other words Lincoln will disinherit Lacy if he does not show himself obedient and dutiful.

85 *the true bias of my love* I.e. 'if you shy away out of the line in which my affection naturally runs'. The metaphor implied here in 'bias' comes from the game of bowls, the bowl being loaded so as to make it move out of ('start from') a straight course.

86 *My Lord, I will...* Lacy's speech does not answer Lincoln directly. He turns the point, but nevertheless appears to be respectful and obedient.

90 *thirty Portuguese* Something over £100, and a generous exchange for 'those words'.

94 *clap swift wings...* add swift wings to your plans.

99 *but I'll o'erreach his policies* 'Policy' had always the sense of crafty Machiavellian designing, and 'o'erreach', with the idea of 'outdoing', 'going one better', is also a very Elizabethan term (cf. *The Overreacher*, the title of a book on Marlowe by H. Levin).

110 *tried itself in higher consequence* been tested in even more important matters.

S.D. *with a piece* Most probably a gun (Rafe is off with the army to France). Some editors give it as a piece of leather.

120 *go to* A common Elizabethan colloquialism, close to the relatively modern 'get away with you', or 'Garn' as Shaw's Eliza puts it.

125 *pishery-pashery* A sort of jovial contempt implied. The term is a favourite one with Eyre, who uses it again in l. 163; these jingly collocations (gibble-gabble, jiggy-joggy, etc.) are fairly common in Dekker.

132–3 *Firke, my fine firking journeyman* To firk, say Skeat and Mayhew, is 'to move about briskly, to frisk, gallop'. It could mean a variety of other things including 'to cheat, or rob' and 'beat or trounce'. The word occurs in this play rather frequently, usually as a pun on the character's name, and was evidently good for a laugh.

132–3 *journeyman* A journeyman had served his apprenticeship and was now qualified to work at his trade for daily wages (*journée*, a day). He was not yet a master (but later, Eyre appoints Hodge, his senior journeyman, as his successor in the shop).

135–6 *the Gentle Craft* See note on the Preface, p. 109.

140 *master cormorant* Firke is the clown of the company and has a fool's licence. So 'cormorant' is a play on 'colonel' (or in Eliza-

bethan spelling 'coronel', which is a little closer in sound). There is also some point in the pun, as the cormorant was a greedy bird of prey, gobbling up others' goods, much as the officer takes good men away from their home, wife and business.

148 *occupied* The word had bawdy overtones. Cf. Doll Tearsheet in *2 Henry IV*: 'God's light, these villains will make the word captain as odious as the word "occupy", which was an excellent good word before it was ill sorted' (II, 4).

150 *the Londoners are pressed* I.e. impressed by the press-gangs which forced men into the army and navy.

156 *Gramercy* From Fr. *grand merci*, great thanks.

163–4 *your pols and your edipols* 'Edepol' was a mild exclamation involving, as its literal sense, the invoking of Pollux. Eyre thinks his wife has too much to say for herself and silences her with this bantering, half-nonsensical satire on the awkward little speech she has made.

164–5 *let your head speak* The emphasis is on 'head' (cf. Cicely Bumtrinket); cf. 'use your loaf'.

166 *Yea, and the horns too, master* This kind of joke is so persistent that its survival is one of the wonders of literature. A man whose wife had been unfaithful to him was a cuckold, and the notion was that he then grew horns on his brow. The origin of the idea is lost in antiquity, but it kept the Middle Ages amused, was apparently a sure success in Elizabethan times, ran throughout Restoration comedy, was killed in England only by the restraints of Victorianism, and turns up in France to this day.

170 *dankish* damp.

171–2 *an hackney to him* a common drudge.

172 *Termagant* A fictitious Mohammedan God, represented in medieval plays as a fiery, violent character.

179 *thou art a gull* A 'gull' is a fool, a gullible person, easy to trick; the 'an' means 'if'.

183 *A common slight regard shall not respect him* He will be accorded something more than the low respect in which common soldiers are generally held.

185 *thou shalt not want* you shall lack nothing.

201 *his pickthank tales* Tales told by this flattering parasite to gain favour.

206 *colours* The standard of the regiment. Hence 'join your unit'.

213   *Peace, you cracked groats, you mustard tokens!*   The same bantering tone of humorous abuse. A groat was a fourpenny piece, so a cracked groat is a worthless article. A token was also a cheap coin, while the phrase 'mustard token' was used of the yellow spots on the body of a victim of the plague. The two meanings together are sufficiently uncomplimentary.

221   *you bombast cotton-candle-quean*   A 'quean' was a prostitute, and the line 'work for your living with a pox to you' continues the mildly bawdy humour of this piece of banter. Bombast is a cotton wool used for stuffing, and cotton-wick candles were sold fairly cheaply by travelling journeymen. Here is a living for Jane which is at any rate more wholesome than Firke's suggestion that in her husband's absence she should 'be doing with me, or my fellow Hodge' (cf. also the phrase as used in II, 3, 41).

231   *the third shall wash our souls at parting*   Firke means that the third twopence will buy them a drink. Literally it might mean that it will buy the Church's blessing on them.

232   *firk the* Basa mon cues   'Basa mon cues' seems to have preceded 'frog' and 'parley-voo' as a term of humorous abuse of the French. The expression is corrupted from 'baisez mon cul' and was quite common in the north of England. 'It was shouted by small boys at anyone of foreign appearance, and so became a kind of generic appellation' (W. J. Halliday).

234   *cram thy slops with French crowns*   Slops were loose, baggy breeches, and the crowns were the coins a soldier would hope to collect as loot.

243   *Made up and pinked*   Pinking was the process of making eyelet-holes in the shoe leather, then to be decorated with gilt or silver buckles.

248   *mo* more.

ACT I. Scene 2

18   *against I shall be Lady of the Harvest*   'against' means 'in preparation for the time when'. In the harvest festivities, a Lady of the Harvest would be chosen (like the May Queen) and would represent the earth's abundance, garlanded with fruits and flowers.

22   *by Doctor's Commons*   A district near St Paul's where church lawyers lived, well-known also as the resort of idlers.

26 *Here 'a wore* 'a = he.

32–3 *'Marry, go up', thought I, 'with a wanion'* 'Go up' probably meant something like 'clear off'. 'Wanion' originates in the waning of the moon and the idea that this is an unlucky time. So 'with a wanion' means something like 'and bad luck to you'.

33–4 *Are you grown humorous?* you're getting moody, are you? A man's 'humour' by this time was 'the kind of person he was', but a 'humorous' man might be an affable character one day and a proud, unapproachable one the next.

39 *stamped crabs* Crushed crab apples, very sour.

40 *sour as verjuice* The acid juice of green (unripe) fruit.

41–2 *much in my gaskins, but nothing in my netherstocks* Gaskins were wide trousers first worn in Gascony; netherstocks were stockings. The phrases are most likely proverbial, meaning 'I will be perfectly polite to you socially but will not take you into my private confidence'.

44 *Go by, Hieronimo, go by* Hieronimo is the principal character in Kyd's revenge play *The Spanish Tragedy*, written *c.* 1589, and by the time of *The Shoemaker's Holiday* something of a joke. The line 'Hieronimo beware, go by, go by' became a catch-phrase.

45–6 *old debts...goose giblets* The general sense is that what you have will balance your losses and it is best simply to accept the change. 'Driblets' are small sums of money. The second line seems proverbial (Dekker uses it again in *Westward Ho*, v, 4, 282: ' and therefore set the Hares-head against the Goose-giblets, put all instruments in tune, and every husband play music upon the lips of his wife, whilst I begin first').

52–3 *let him go snick-up, young mistress* let him go hang; said to have survived into this century as a north-country expression with this same meaning (see *English Dialect Dictionary*, ed. Wright).

57 *My cambrick apron* 'A kind of fine white linen originally made at Cambray in Flanders' (*O.E.D.*). 'Stomacher', 'an ornamental covering for the chest (often covered with jewels) formerly worn by women under the lacing of the bodice' (*O.E.D.*).

64 *have at up tails all* The general sense is 'get the horse moving and be off', but 'up tails all' was also the name of a game as well as the refrain of a song, so it serves generally to communicate the high spirits of the occasion.

ACT I. SCENE 3

1 *shapes* transformations. Lacy is thinking comically that if he cuts a strange figure as Hans Meulter, a Dutch shoemaker, at least he is in good company, for the Greek gods sometimes went wooing in still more grotesque disguises (e.g. Jupiter who came to Leda as a swan, and to Europa as a bull).

ACT I. SCENE 4

2 *the fat brewis of my bounty* 'Brewis' is meat broth or bread soaked in broth or dripping.

3–4 *to see my walks cleansed* to sweep the pavement outside the shop.

4–5 *you powder-beef queans* Powder-beef had been sprinkled with salt, i.e. salted beef. 'Queans' is the familiar colloquialism (cf. note on I, 1, 221), probably less insulting in this sort of phrase than 'slut' would be today.

5 *Madge Mumblecrust* A character in *Ralph Roister Doister*, the popular comedy (*c.* 1553) by Nicholas Udall.

7 *kennels* gutters.

10 *bandog and bedlam* Speaking furiously as a chained mastiff and wildly as a lunatic.

18 *a souse wife* One who would wash and pickle a pig's trotters, etc. But 'souse' is a general word meaning immerse or soak in water.

23 *towards* starting.

38 *want* lack, i.e. he will thrash her if she has forgotten to provide it.

38–9 *I'll swinge her in a stirrup* Swinge means 'to thrash', and the stirrup was a shoemaker's leather, evidently often used for this purpose as the verb 'to stirrup', surviving into the eighteenth century, meant just that.

38–9 *Lacy's song* 'There was a farmer of Gelderland, Jolly they are; He was so drunk he could not stand, Druken they are. Clink once the cannikin, Drink pretty mannikin' (all Lacy's 'Dutch' speeches are translated literally and not idiomatically in these notes).

45 *Upsolce* Like much else, this is neither Dutch nor English, but there is an association in the phrase 'upsey-Dutch' which meant 'in the Dutch fashion' and was often combined with the idea of heavy drinking (e.g. 'drink me upsey-Dutch. Frolic and fear not', *Beggar's Bush*, III, 1, 3).

49   *Saint Hugh's bones*   When St Hugh's body was taken down from the gibbet by his fellow shoemakers, his bones were turned into tools for the trade (see pp. 136–7 above).

50   *uplandish*   foreign.

53   *a hard world!*   This is what employers say to men looking for work ('It's a hard world, and I've got enough men as it is...').

58   *entertain every butterbox*   every Dutchman—Eyre's wife is speaking contemptuously (cf. II, 3,143–5: 'they may well be called butter-boxes, when they drink fat veal, and thick beer too'). The Dutch were proverbially great drunkards (cf. 'Dutch courage').

75   *gallimaufry*   A hot pot of various meats, hence a hotch-potch of odds and ends, often used contemptuously.

78   *Goeden dach, meester, ende u vro oak*   'Good day, master, and your wife too'.

79   *Nails!*   God's nails, an oath used several times by Firke.
     If I should speak after him without drinking   I couldn't repeat his barbarous lingo without having a drink first.

80   *friend Oak*   Firke thinks (or pretends to think) that Hans' last word (oak) is his name. It really means 'also' (cf. German *auch*).

89   *Yaw yaw..*   'Yes, yes, be not afraid. I have all the things to make shoes great and small' ('Clean', cf. *klein*).

93–4   *the mystery of Cordwainers*   The trade of the cordwainers, leather workers. The Worshipful Company of the Cordwainers is still a City Company in London (see Introduction, p. 6).

95   *Ik weet niet...*   'I know not what you say; I understand you not.'

96   *Why, thus man*   Firke presumably demonstrates in dumb-show, and his 'ich verstaw you niet, quoth 'a' is humorously disparaging.

108   *your Trullibubs*   Here meaning a slut, but more generally something worthless, often used of a cheap food like tripe. Various forms of the word existed, most commonly 'trullibub', sometimes 'trollibags', often in the phrase 'tripe and trollibobs'.

110   *Use thyself friendly*   act in a friendly way.

113   *Gargantua*   A giant of prodigious appetite in Rabelais' *Gargantua and Pantagruel*.

116   *O, ich wersto you...*   'Oh, I understand you: I must pay for half-a-dozen cans. Here, boy, take this shilling. Tap one freely.'

120   *Come, my last of the fives*   Eyre is calling the boy and making a

pun on 'last'. The number-five shoemaker's last would be for a pair of very small shoes.

129 *Dame Clapper-dudgeon* 'A *clapperdudgeon* is, in English, a beggar born' (Dekker's *Villainies Discovered*), one in a catalogue of colloquial terms distinguishing various types in what he calls 'the regiment of rogues'. The beggars carried a clap-dish to gain attention, and Mistress Eyre's interruption of the drinking made a similarly insistent and unwelcome sound.

131 *soused conger* pickled conger-eel.

131 *mad Hyperboreans* 'Hyperborean' is a term for inhabitants of the extreme north. Eyre is constantly giving his men these names with the jocular abusiveness schoolmasters sometimes favour.

137 *to go behind you* The shoemakers were keen on rules of precedence. An earlier example is Firke's reminder that senior journeymen have the right to the first drink (l. 102).

ACT II. Scene 1

1 *brake* Clump of bushes, or thicket.

6 *take soil* The wild boar would take refuge in a pool or some muddy place.

7 *embossed* blown and fatigued with being chased. (Cf. 'The boar of Thessaly was never so emboss'd', *Antony and Cleopatra*, IV, 2, 3).

13 *pale* fence.

ACT II. Scene 2

2 *Upon some, no! Forester, go by!* 'Upon some' is Sybil's catch-phrase (she has it again ll. 16 and 28), as 'but let that pass' is Mistress Eyre's. A variant on 'upon my word', it has no apparent modern equivalent. Sybil's disclaimer would probably go into modern speech as: 'For heaven's sake, no! Forester? Get away with you!'

17 *Wounds!* An oath (like Firke's 'Nails' in the first act).

22 *his horns will guide you right* Again the play on 'horns' (the deer's, and also the mark of a cuckold and a fool, cf. note on I, 1, 166). Warner accepts this as humorously insulting, hence his half-appreciative 'Th'art a mad wench': mad as in madcap.

25 *places of resort* inhabited places.

28 *good honey-sops* sops of bread and honey.

31 *a deer more dear* The Elizabethans liked puns, and this is just as

well, for word-play on 'dear/deer' and 'heart/hart' is kept going until the arrival of the Lord Mayor (and even then, Hammon hears a double-entendre in the Mayor's line: 'I hear you had ill luck and lost your game').

ACT II. SCENE 3

1   *Ick sal yow wat seggen...* 'I'll tell you what, Hans. This ship that comes from Candia [formerly Crete] is all full, by God's sacrament, of sugar, civet, almonds, cambrick and all things, a thousand thousand things. Take it, Hans, take it for your master. There are the bills of lading. Your master, Simon Eyre, shall have a good bargain. What say you, Hans?'

7   *Wat seggen de reggen* No sense here. Firke is being facetious, mimicking the language as a kind of baby-talk. Hodge, the melancholy one, then needs prodding into amusement.

9   *Mine liever broder, Firke...* 'My dear brother Firke, bring Master Eyre to the sign of the Swan. There shall you find the skipper and me. What say you, brother Firke? Do it, Hodge. Come, skipper.'

16 *a bable* a trifle, a play-thing ('bauble' is a common alternative form).

24 *earnest penny* a down-payment.

26–7 *Saint Mary Overy's bells* The Church of St Mary Overy (now the Cathedral of St Saviour) in Southwark.

30 *Monday's our holiday* Cobblers' Monday (there was a long-surviving tradition that cobblers' shops remained closed on Monday).

43 *I'll firke you* beat you.

52 *sort* pack.

54 *a venentory* inventory (another Malapropism).

57 *more maids than Mawkin* Mawkin or malkin, little moll, an untidy kitchen girl; a proverbial variant on the 'more than one pebble on the beach' theme.

66 *tannikin* 'A diminutive pet-form of the name Anna, used especially for a German or Dutch girl' (Skeat). The epithet 'brown bread' adds a suggestion of coarseness (white bread being more refined).

67 *Move me not!* don't make me angry.

74   *chitterling* 'Smaller intestines of the pig etc., especially when fired or boiled' (Skeat).

82   *Where be the odd ten?* Eyre pretends for a moment to wonder where the other ten cans are.

    Madge Eyre seems to use this as a diminutive of Margery, his wife's name.

90   *yark* To draw the stitches tight.

91   *let me alone, an I come to't* there's no one to beat me if I am really put to it.

92   *from the bias* beside the point.

101   *Skellum Skanderbag* There seems to be no particular point in the reference to Iskander-Bey, which was the Turkish name for Castriota, the fifteenth-century Albanian patriot. It pairs alliteratively with 'skellum' (a rascal, cf. Dutch 'schelm', a rogue), and that is presumably enough.

102   *silk Cyprus* Cyprus and Candia, the islands from which the silk and sugar have come. It is part of Firke's role as a comic to wrap his sensible notions up in nonsense.

104–5   *a guarded gown* Ornamented round the borders.

108   *Silk and satin!* Eyre's high spirits are intensified by the fun of dressing up and playing a part. By pretending to be an Alderman he will get the articles on credit.

110   *beaten damask* 'orig. hammered; hence overlaid or inlaid; embroidered' (Skeat).

111   *rearing of the nap* Firke is helping rather too vigorously. Eyre tells him to go carefully ('softly') or he will ruffle the nap (the threads of cloth at the surface, the pile) and make the clothes look worn out.

115   *give you the wall...* give precedence to him, and address him as 'Right Worshipful'.

120   *brisk* spruce.

127   *Godden day...* 'Good day, master. This is the skipper that has the ship of merchandise. The commodity is good; take it, master, take it.'

132   *De skip...* 'The ship is on the river. There are sugar, civet, almonds, cambrick, and a thousand thousand things. God's sacrament, take it, master; you shall have a good bargain.'

136   *carrot roots, turnips* Both were introduced into England from Holland, the carrot in the sixteenth century, and the turnip (as a

home-grown vegetable) not till near the end of the seventeenth. It seems odd to us that such homely articles should be included in this kind of list, but at the time they were evidently still exotic vegetables.

140 *Yaw yaw...* Yes, yes, I have drunk a lot.

146 *perdy* A mild oath (cf. Par Dieu).

ACT III. Scene 1

20 *square* quarrel.

22 *strange in fancying me* 'Strange' sometimes meant remote or cold in manner.

24 *so fond to fond* so foolish as to bestow my love ('fond', the verb, is 'to found', its form being modified for the sake of the word-play).

25 *quit it* requite it.

45 *tourney and at tilt* A tourney or tournament; a tilt was a combat between two armed men on horseback and with lances. The lady's glove would be worn as a pledge of the knight's devotion. Hammon is very much the modern young man and these chivalric traditions are all absurd, old-fashioned stuff to him.

51 *the Old Change* the old Exchange.

56 *mammet* A variant of maumet, from Mohammed, hence a false god or idol, the term then coming to mean a doll, a puppet, as applied abusively to a girl. In similar circumstances Capulet uses it in his outburst against Juliet:

> And then to have a wretched puling fool,
> A whining mammet in her fortune's tender
> To answer 'I'll not wed', 'I cannot love'.
>
> (*Romeo and Juliet*, III, 5, 185–7)

75 *it likes yourself* it pleases you.

95 *start him* find him, flush him out of cover.

97 *a dozen angels* An angel was a gold coin with the Archangel Michael and the Dragon as its device. Its value is given as about 10s. at this time, but that obviously does not mean very much unless related to value in spending power.

ACT III. Scene 2

8   *too compendious* This should mean 'economical' (a compendium being an abridgement or a summary), but presumably Mistress Eyre intends the opposite, as 'tedious' generally carried the specific sense of 'long-winded'. She is, of course, showing off the big word she has picked up, and in his next sentence Firke satirises her for it.

12   *Nay, when?* I.e. 'when are you going?' (an impatient exclamation which could also be in Elizabethan a cry for attendance).

13–14   *that humour* Melancholy was one of the four humours, the others being choler, phlegm and blood. The preponderance of one of these, thought of as fluids in the body, was believed to determine one's character or mood.

22   *Me tank you, vro* I thank you, mistress.

27   *your back friend* A backward, or perhaps deceptive, friend. Also no doubt having, in its context, a faintly bawdy flirtatious double meaning.

29   *Yaw, ic sall, vro* Yes, I will, mistress.

35   *farthingale-maker* A farthingale was a petticoat made to project around the waist by whalebone or hoops. The skirt was worn over this.

36   *French-hood* These had large flaps standing out alongside the temple. Hodge's aside about the pillory is suggested by this, the flap resembling the pillory boards, and the face, in both instances, coming between them.

43   *Gracious street* Gracechurch Street.

48   *a fan or else a mask* The fan was new to England in Elizabeth's time. 'It was worn hanging from the point of the stomacher: it often contained a small mirror' (*Shakespeare's England*, vol. 2, p. 97). 'Masks were of various colours, and were much worn by ladies of quality when riding. The eyeholes at times were filled with glass' (*ibid.*). The mask guarded the complexion from sunburn, thought to be unsightly and low class.

54   *Ich bin vrolick...* I am merry: let me see you so.

55   *drink a pipe of Tobacco* The usual Elizabethan expression (cf. Thos. Pennant *c.* 1760: 'The first who smoked or (as they called it) drank tobacco publickly').

69   *impotent* injured.

89   *more stately than became her*   I.e. she put on airs.

92   *Bye nor bah*   The general sense is that she left without a word (presumably without a polite goodbye, or anything that might have shown there was a disagreement between them).

       ka me, ka thee   help me and I'll help you. Skeat gives 'Kae me and I'll kae thee' as a still-surviving Scottish expression.

107–8   *brave and neat*   well-dressed and trim.

116   *smug up your looks*   cheer up, smarten up.

119   *this famous year now to come*   Historically 1436, but perhaps Dekker sees it in his present time when the 'year now to come' would be the first of the new century.

126   *Yaw. My mester...*   Yes, my master is the great man, the Sheriff.

133   *I smell the Rose*   On the threepenny coin the Queen was shown in profile, a rose behind her ear (this was so on the sixpenny and threefarthing pieces also). The threepenny piece was the coin used as Maundy money, distributed by the almoner, and it was not in general use. Mistress Eyre is living up to the grandeur of her new position in bestowing this so regally upon Firke.

134–5   *do not speak so pulingly*   'Pulingly' suggests the whining of a child, but presumably the Mistress has been using a grand lady's voice which Hodge finds la-di-da. Firke too prefers the less refined notes of the 'old key': and as he urges a return to speech 'with a full mouth' we assume that the affected tones of polite society tended to be mincing and restricted, then as now.

141   *See, mine liever broder*   See, my dear brother, here comes my master.

147   *flap of a shoulder of mutton*   Trimmings of sheep's wool; probably also a reference to the flaps of head-dresses, as described in the note to l. 36 above.

150   *an hundred for twenty*   For the recent piece of business between himself and the Dutch skipper, Eyre tells Hans he shall have a hundred Portuguese (coins) in return for the twenty he had given as earnest (see II, 3, 20).

162–3   *some morris*   Morris dancing was for festive occasions, notably May-day, and involved dressing in fancy costume to represent legendary characters such as those in the Robin Hood stories. Eyre wants a suitable entertainment: dancing, music ('some odd crotchets'), and an appropriate tableau, emblem or design ('some device').

ACT III. Scene 3

26   *sack and sugar*   Sack was a dry white wine (originally 'wyne seck', *vin sec*) imported from Spain or the Canaries. It was popular in Elizabethan times (cf. Falstaff: 'If sack and sugar be a fault, God help the wicked', *1 Henry IV*, II, 4). Compare, too, the eighteenth-century song *Simon the Cellarer*: 'Of sack and canary he never doth fail' (that Simon also had a woman called Margery about the house, and was a jolly fellow—one wonders if the legends of Simon Eyre merged into the character of this other Simon). From this sentence we gather sack and sugar to have been regarded as a drink for old-timers.

36   *The ape still crosseth me*   This seems rather fierce, but the comparatively modern father may refer to his daughter as a 'monkey' without meaning anything particularly insulting.

39   *fine cockney*   Perhaps simply meaning a Londoner (it is the Lord Mayor talking), but also possibly carrying one of the archaic meanings of the word: 'an egg: or perhaps one of the small or malformed eggs called popularly "cock's eggs..." (also) "A child that sucketh long"' (*O.E.D.*).

40   *prove a cockscomb*   make a fool of yourself (by wanting to marry above the proper station).

44   *Wash*   Eyre's contemptuous tone dismisses the 'courtier' as a sort of half-man, 'the waste water discharged after washing', 'swill for swine', 'stale urine'—there are plenty of uncomplimentary uses of the word to choose from without resorting to the usual suggestion that it is an abbreviation of 'washical' ('What shall I call').

50–1   *he should pack*   I should send him packing.

66   *Ic be dank, good frister*   I thank you, good maid.

74   *Stratford Bow*   Stratford-le-Bow (or Stratford-atte-Bowe in Chaucer's time) was still outside London, beyond Whitechapel, in the country where one could buy cakes and cream at farms.

ACT III. Scene 4

9   *coy*   disdainful, with perhaps the sense of 'playing hard to get' (the sense of the word as used in l. 65 below will not fit here, for Rose was very far from feeding him with 'sun-shine smiles, and wanton looks').

10   *curious*   scrupulous, punctilious.

22   *What is't you lack*   Jane is crying her wares (cf. Orlando Gibbons' *Cries of London*: 'What is't ye lacke? Fine wrought shirts or smocks').

25   *How do you sell*   how much does it cost?

46   *woman's fray*   the sort of opposition to be expected from a woman.

59   *silly*   Here meaning 'a poor sort of conquest'.

95   *bill*   document.

115 *rude*   Hammon threatens to become rough, or like Chaucer's Troilus to 'mak it tough'.

ACT IV. SCENE I

1   *Hey down, a-down*   though Hodge congratulates them with the words 'well said', one presumes that they are singing (cf. *The second Three-mans Song* which has similar words in the third verse).

11   *Forware Firke...*   'Indeed, Firke! Thou art a jolly youngster. Hear me, master. I pray you cut me one pair of vamps for master Jefferies' boots.' The vamp is the part of the shoe covering the front of the foot.

18   *counterfeits*   Repairing the shoes by putting on new vamps is a sort of patching or 'counterfeit'. The word-play then takes up the sense of false coinage, which would 'not pass current'.

22   *aunts*   harlots (hence the bawdy double-talk of Firke, l. 26 below and following).

29   *this gear will not hold*   Various possible meanings here. Firke is at work on a pair of shoes, chatting and singing while he does his job. So the phrase might be said frowning at the repair which doesn't look like lasting. Or it might be said, as has been suggested, in irritation at the silly bit of song he has just sung (this seems unlikely). It could also, in an equivocal fashion, play on the theme of the bawdy talk, for 'gear' could mean 'organs of generation' (*O.E.D.*) and Firke has just been making tasteless jokes at the expense of the war-wounded Rafe.

33   *Roger Oatmeal*   Cf. 'Roaring boys and oatmeals'. The phrase occurs in Ford's *Sun's Darling* (I, 1) and Firke's term is directed to the 'melancholy Hodge', ironically characterising him as a roisterer

and a blood. The precise meaning of the joke about bagpuddings
(which were simply puddings boiled in a bag) is anybody's guess.

58 *Vat begaie you* What do you want? What would you, girl?

61 *Vare ben your edle fro?* Where is your noble lady? Where is your
mistress?

62 *London house in Cornwall* Either Sybil or the old text is mixed-
up; she means Cornhill.

64 *I stand upon needles* I am all impatience (like a cat on hot bricks,
anxious to be off).

67 *budget* A wallet or pouch, usually of leather. So 'in my budget'
is a near-equivalent to 'up my sleeve' with (possibly) an additional
bawdy suggestion following the 'double entendre' of Hodge's
little joke.

68 *Yaw, yaw...* Yes, yes, I shall go with you.

69 *make haste again* hurry back.

ACT IV. Scene 2

2 *mass* A mild oath (by the Holy Mass).

23 *the sign of the Golden Ball in Watling Street* All shops had their
painted signs. Eyre's own shop is 'at the sign of the last in Tower
Street'.

32 *Saint Faith's Church under Paul's* 'At the west end of this Jesus
chapel, under the choir of Paul's, also was a parish church of St
Faith, commonly called St Faith under Paul's, which served for
the stationers and others dwelling in Paul's churchyard, Paternoster
road and the places near adjoining' (Stow).

40 *'Snails* 'His nails'—i.e. the nails with which Christ was crucified
—an oath common in medieval and Tudor times.

43 *A thing?* Firke turns most remarks to bawdry, here playing
on the slang connotation by which the 'private parts' were
'things'.

51 *I hold my life* as sure as I live.

53 *murrain* A plague of cattle (i.e. 'How the plague...').

54 *ague fit* Presumably because Rafe is shaking and jumping up
and down in his excitement.

70–1 *Cripplegates* 'Cripplegate, a place, saith mine author (John
Lydgate) so called of cripples begging there: at which gate it was
said, the body (of King Edmond the Martyr) entering, miracles

were wrought, as some of the lame to go upright, praising God'
(Stow, quoted by W. J. Halliday, who points out that this was the
site of the Fortune Theatre where *The Shoemaker's Holiday* may
have been produced).

72–3 *wedding and hanging* Cf. *Merchant of Venice*, II, 9, 82–3:

> The ancient saying is no heresy,
> Hanging and wiving goes by destiny.

ACT IV. Scene 3

31 *Forware, metress* Indeed, mistress, tis a good shoe; it shall do
well or you shall not pay.

37–8 *Yaw, yaw* ... 'Yes, yes, I know that well. Indeed, 'tis a good
shoe: 'tis made of neat's leather; just see, my lord.'

45 *make things handsome* tidy up and see that everything is looking
its best.

ACT IV. Scene 4

7 *had given head* enlarged (i.e. had given the attempt its head).

14 *and he hath done so* The sense may not be obvious here. The
meaning is: 'I love your nephew Lacy too dearly to wrong his
honour so much; and the person who has done him this wrong is
the one that first advised him not to go to the war in France'.

27 *I'll straight embark for France* 'Embark' is transitive, i.e. 'If I
find him, I'll have him shipped off to France immediately'.

48 *honnikin* Clearly contemptuous. Skeat links the word with the
German 'hohn' meaning scorn, derision, and with the Middle
High German word 'hone' meaning 'a despised person' (Skeat
and Mayhew, p. 198).

56 *That's brave!* An ironical exclamation, something like 'that's
just marvellous', except that 'brave' also carries the idea of
grandeur, doing things in the grand style.

74 *Syb, your young mistress* ... An aside to Sybil, meaning 'I'll
fool ("bob") them about her'. A 'bobber' was a slang term for a
cheat.

77–8 *no maw to this gear* Firke is being difficult. He knows what
Lincoln and the Mayor want to get out of him, but uses his comic-
man's licence to evade their questions. Here he is pretending that
he thinks they are trying to marry him off to Sybil for whom he

says he has no taste. 'Maw' = stomach, i.e. inclination. 'Gear' (in this context) = dress ('bit of stuff').

81   *the tune of Rogero* Firke has told them he is 'lusty Roger's chief lusty journeyman' (ll. 61–2). He is still answering in the quaint, allusive manner of the accepted Clown; and here the reference is to a well-known dance tune, called the Rogero. Jack Slime proposes it in *A Woman killed with Kindness*: 'I come to dance, not to quarrel. Come, what shall it be? Rogero?' (I, 2, 32).

87–8   *the shaking of the sheets* There is plenty of method in Firke's antic disposition. He fools his questioners now with a joke which is dangerously near to their secret fears. 'The shaking of the sheets' is an old dance (cf. Eyre, later in the play; 'mark this old wench, my King, I danced the shaking of the sheets with her six and thirty years ago', v, 5, 30–1), but it also means that they are getting married, making the nuptial bed.

88   *I'll so gull these diggers!* I'll fool these people who are trying so hard to dig information out of me. An aside, perhaps to Sybil, out of the corner of the mouth.

91   *Canst thou? In sadness?* Being serious now, can you tell us?

94   *what I'll bestow of thee* What I will give you in return.

97–8   *aurium tenus…genuum tenus* The Latin means 'as far as the ears…as far as the knees'. So the sense of the passage is this: 'Give me ten gold pieces and I will go the whole way (up to the ears), but if you can only manage silver, then you can have only half of my information (up to the knees).' There is probably a joke in the 'tenus…ten', and perhaps in the 'aurium' and 'gold' (='aurum' in Latin).

99   *a new pair of stretchers* 'Stretchers' are shoe-stretchers; here an apt metaphor, for the shoemaker is stretching the truth. Meaning: 'give me sufficient gold or silver and I will serve you—with more lies'.

100   *an angel* A gold coin (see note on III, 1, 97).

103–4   *my corporation* a member of my craft-guild.

104   *firked and yerked* The general sense is clear enough. As usual, one wonders which of the dictionary definitions, if any, is intended. Probably 'hard-pressed, driven' ('This shall serve to firke your adversary from court to court', Brome, 1640). 'Yerked' is a technical term meaning to draw stitches tight (as in 'yark and seam', II, 3, 90).

110 *Hans-prans* A piece of rhyming slang.

112 *by this rush* Probably an allusion to rush rings, sometimes used
for weddings, often by a bridegroom who had no intention of being
held by his vows. Used at this juncture, it is like Touchstone's
'oath referential', and the roguish undertone is well in character.

116 *London Stone* This marked the meeting place, in London, of
the great Roman roads. The stone has been encased in the wall of
St Swithin's Church, Cannon Street, a few yards from its place of
origin.

  Pissing Conduit A small stone water-cistern near the present
site of Mansion House. Presumably it leaked, Firke's reference to
Mother Bunch, the famous hostess of a tavern, meaning that there
was little to choose between this and the watery ale served at her
establishment. Jack Cade, in *2 Henry VI*, IV, 6, 3, says: 'And here
sitting upon London Stone, I charge and command that, of the
City's cost, the pissing conduit run nothing but claret wine this
first of my reign.'

121 *a swearing church* A church whose name could be the subject
of an oath ('by my faith').

125 *incony* The term appears several times in plays around the
turn of the century and then quickly disappears from the language
altogether. The sense here may be 'incognito' (or French 'incon-
nu') which makes sense by derivation; but the meaning in most
other contexts seems merely to suggest something pleasant or
'nice' (e.g. 'It makes you have, O, a most incony body', Middleton).

130–1 *No spirit* Lincoln has said 'my nephew Lacy walks in the
disguise of this Dutch shoemaker'. 'Walks' is what a ghost does;
hence Firke's 'no spirit'.

140 *hey pass, and repass* A term used in conjuring, here meaning
'when they are up to their tricks' (of getting married). 'Pindy
pandy, which hand will you have' refers to the children's game of
handy-dandy, this coming to mind because of the giving of hands
in marriage.

153–4 *coney-catched* caught by trickery (it was a popular term for
a decade or so after 1591 when Greene published his *A Notable
Discovery of Coosnage. Now practised by sundry lewd persons, called
Connie-catchers and Crosse-biters*).

157 *at the Savoy* The Savoy Palace possessed the right of sanctuary,
and the chapel of the hospital served as a parish church.

162–3 *at the Woolsack in Ivy Lane* The Woolsack was a popular low tavern.

165 *Alack, alack!* Probably a snatch of song, suggesting that young women should stand firm ('hold out tack'), for maidenheads ('smocks' being a euphemism) are in danger ('go to wrack') in the general excitement ('This jumbling').

ACT V. SCENE I

1 *Stay my bully* No modern equivalent of 'bully' suggests itself. 'Comrade', 'chum' perhaps: the sort of hearty term of back-slapping endearment that we now tend to avoid.

14 *vah* A contemptuous exclamation, like the Latin 'vah' or 'vae', as in 'vae victis'.

32 *marchpane* marzipan.

42 *careful* worrying.

44–5 *to see my new buildings* 'Simon Eyre builded the Leaden Hall for a common garner of corn' (Stow) (see p. 139).

49 *Cappidosians* Caperdewsy and Caperdochy were terms for prison and the stocks. This seems to be the likely reference, in key with the rest of Eyre's banter. Skeat, however, thinks it 'is evidently a jocose expression for mad-caps, with a punning reference to the...special headgear of the London apprentices'. There is no reason why it should not be both.

49 *served at the Conduit* 'Anciently it was the general use and custom of all apprentices in London...to carry water tankards to serve their masters' houses with water, fetched either from the Thames or the common conduits of London' (Stow, p. 329, quoted by W. J. Halliday). See also v, 5, 181–2, where Eyre says that he himself 'in time past...bare the water tankard'.

50–1 *I would feast them all* See pp. 138–9.

53 *upon every Shrove-Tuesday* Apprentices traditionally enjoyed a holiday on Shrove-Tuesday, but there seems to be no historical evidence for making Simon Eyre the originator.

54 *the pancake bell* This was another traditional feature of the day. It was becoming a thing of the past in the eighteenth century, but is said to have survived throughout the nineteenth in Buckingham-shire (see F. O. Mann's edition of Deloney's works, p. 532).

ACT V. SCENE 2

4 *Were Hammon a king of spades* Not (primarily) the card, but the implement. So the general sense is: 'However great a personage Hammon may be, he shan't interfere in your business and get away with it.' 'Delve in thy close' (dig in your allotment) is a homely way of referring to Rafe's rights over Jane, on which Hammon is trespassing. 'Sufferance' could then mean 'the suffering of a penalty'.

18–19 *neither Hammon nor hangman* A double pun: (1) the similar sound; (2) the Hammon in the play and the Haman in the Book of Esther, who having prepared a gallows fifty cubits high for his enemy was strung up on it himself.

31–2 *cry clubs for prentices* 'Cry clubs' was the rallying cry for groups, especially apprentices, who wanted to start a street-fight. In the first scene of *Romeo and Juliet*, for instance, the citizens cry out 'Clubs, bills and partisans' (l. 74) as the fracas begins between the supporters of the two houses.

35 *bird-spits* I.e. knives or swords.

57 *mend it* alter it (if you can).

61 *affect* love, have affection for.

63 *these humble weeds* clothes (the contrast between the courtly gentleman in his fine wedding rig-out, and the war-wounded Rafe in his plain shoemaker's dress, must have been touching to the girl, put between them and looking from one to the other).

71 *a busk-point* The 'point' was a piece of lace by which the whalebone stiffening of the bodice could be tied down into position.

75 *Bluecoat* A reference to the servant's blue livery.

76 *Saint George's Day* Firke's threats are made in oblique style. On St George's Day the contracts of serving men might be changed and renewed, and so they would have a new livery. This particularly vocal servant, Firke suggests, will find himself with a new (or unrecognisable) livery on this Shrove Tuesday, if he doesn't go carefully.

79 *to clouts* to rags.

91–2 *commodity* A piece of business conducted out of self-interest.

103 *creature* In the vocative (with the sense that the servant is no more than a puppet of Hammon's).

153 *mean* of low birth.

156 *laced mutton* a courtesan (cf. *Two Gentlemen of Verona*, I, I, 102:
'and she, a laced mutton, gave me, a lost mutton, nothing for my
labour'). Not that the phrase applies to Jane: it is rather an incite-
ment to Lincoln, coming as an answer to his question about Lacy.

178–9 *lammed them with flouts* 'Lammed' has the same sense as our
'lammed into them', i.e. attacked them. 'Flouts' means 'mockery
at their expense'.

184 *swagger* One of those expressions that are in vogue for a
decade or so and never mean quite the same again. A 'swaggerer'
was a roaring boy, 'one of the lads' we might say, but in an
aggressive, colourful style. Hodge proposes to 'swagger', to cut
a dash, or as our decade might put it, 'live it up'.

197 *the great new hall* Leadenhall (see note, ll. 44–5). Gracious Street
is Gracechurch Street.

207 *brewis* meat broth (see note on I, 4, 2).

208 *dry fats* dry vats.

210 *collops and eggs* egg fried on bacon (the day before Shrove
Tuesday was called Collop Monday).
  scuttles dishes, platters.

ACT V. SCENE 3

10 *I am with child* I cannot wait.
  huff-cap Skeat quotes from Harrison's *Description of England*:
'Such headie ale and beere as for the mightinesse thereof...is
commonlie called huffecap.' Here it is applied metaphorically to
Eyre's frothy, ebullient character.

ACT V. SCENE 4

11 *rip, knaves* Rip up the purses he has just referred to, or simply
'let rip'.

17–18 *carouse me fathom healths* An invitation to drink deep (cf.
Mercutio's Queen Mab speech where he says the soldier may
dream 'of healths five fathom deep' (*Romeo and Juliet*, I, 4, 85).

25–6 *beleaguer the shambles* besiege the meat-market (Eastcheap
being the great centre).

52 *Islington whitepot* A whitepot was 'a dish made of milk, eggs,
sugar etc.' (Skeat). W. J. Halliday adds that it is shaped like a

hopper (or the vat for the infusion of hops); hence 'happerarse' ('hopper-hipped' and 'hopper-rumped' being other terms in Elizabethan use).

54 *carbonado* Meat or fish which has been cut across, then to be broiled on coals.

54 *avoid, Mephostophilus!* A reference to Marlowe's *Doctor Faustus*, presumably spoken melodramatically as Faustus tries to banish the devil from his presence.

55 *learn to speak of you* Is Simon Eyre going to take lessons in speaking from you?

56 *Mother Miniver-Cap* Mistress Eyre's headpiece for the grand occasion was lined and trimmed with ermine (miniver).

57 *partlets* Linen worn round the neck with a small ruff attached.
   pishery-pashery In I, I, 125 and 163 this is a contemptuous term dismissing the chatter of Firke and Mistress Eyre. But Eyre, like Humpty Dumpty, makes his favourite words mean what he wants them to, and here it applies to dresses and women's matter generally.

57 *flues* Meaning the flap of the hood or skirt; literally 'the large chaps of a deep-mouthed hound' (Skeat).

58 *whirligigs* Toys (like a spinning top), probably extended here to the large, round skirts.
   Rub! Out of mine alley! The expression is from the game of bowls, where the 'rub' is an obstruction caused by an uneven ground, and the bowling 'alley' the place where the game is played.

59–60 *to Sultan Soliman, to Tamburlaine* Great tyrannical potentates of the east, both dramatised by popular Elizabethan playwrights (Kyd's *Soliman and Perseda*, and Marlowe's *Tamburlaine*).

62–3 *frolic free-booters* A free-booter is an adventurer, plundering, like a pirate, for whatever he can lay hands on.

ACT V. Scene 5

1 *the fact* the deed.

16 *Dioclesian* The Emperor Diocletian has been remembered principally for his persecution of the Christians at the beginning of the fourth century, so Eyre's appellation is hardly making his King a compliment. But he has the mad-cap's licence, and so many of his words are chosen for sound rather than sense, that it seems to pass.

16 *Then hump!* A favourite exclamation with Eyre, perhaps accompanied by some comical action (cf. 'cry humpe', ll. 28 and 36 below).

18 *pie* magpie.

23–4 *Tamar Cham* Tamburlaine, the great Cham, an old form of Khan of Tartary (cf. 'fetch you a hair off the Great Cham's beard', *Much Ado about Nothing*, II, 1, 279).

24 *to't* compared to it.

25 *stuff tennis balls* Again cf. *Much Ado about Nothing*: 'No, but the barber's man hath been with him; and the old ornament of his cheek hath already stuffed tennis-balls' (III, 2, 45–7). Dog's hair was more generally used.

30–1 *shaking of the sheets* See note on IV, 4, 87–8.

49 *degenerous* ignoble.

93 *Fair maid, this bridegroom* The nonsense stands in all six quartos. It may, of course, be intentional. 'Is not this banter?' asks W. J. Halliday, and so indeed it probably is, the King having caught on to the general facetiousness.

95 *Then must my heart be eased* The emphasis falls on 'my'. 'Now that Lincoln and Oatley have enjoyed themselves, it is my turn to be gratified.'

136 *The Leaden hall* See p. 139.

162 *buy and sell leather* See Introduction, p. 6.

164 *patent* The authority of the King.

182 *the water tankard* The prentices would be required to carry water from the conduits to their masters' houses (see note to V, 1, 49).

A NOTE ON THE
GENTLE CRAFT, ST HUGH'S BONES
AND SIMON EYRE

Dekker's source was a story-book called *The Gentle Craft*, by Thomas Deloney.[1] This was, as the title-page said, 'A discourse containing many matters of delight, very pleasant to be read, showing what famous men have been shoemakers in time past in this land, with their worthy deeds and great hospitality...declaring the cause why it is called the Gentle Craft, and also how the proverb first grew: "A Shoemaker's son is a prince born".' We also learn how St Hugh became the patron saint of shoemakers, and why it was that 'St Hugh's bones' became a proverbial expression for the shoemaker's tools.

The story goes that Sir Hugh, a king's son, loved a well-born maiden called Winifred. She did not care for him, however, and prevaricated rather in the manner of Rose with Hammon early in *The Shoemaker's Holiday*. Sir Hugh left England greatly discouraged, travelled to Paris, 'counting it the most pernicious place in the whole country', encountered storms and wild beasts, and eventually arrived back 'at a place called Harwich, where for want of money he greatly lamented'. A journeyman-shoemaker befriended him and together they 'agreed to travel in the country'. Winifred in the meantime had become a religious recluse and was imprisoned in Flintshire under the anti-Christian laws of the time. When Hugh learnt this he got himself put in the same prison: 'But during the time that they both lay in prison, the journeyman

[1] *Works of Thomas Deloney*, ed. F. O. Mann (Oxford, 1912).

shoemakers never left him, but yielded him great relief continually so that he wanted nothing that was necessary for him, in requital of which kindness he called them Gentlemen of the Gentle Craft, and a few days before his death, he made this song in their due commendations:

> Of Craft and Craftsmen, more and less,
> The Gentle Craft I must commend;
> Whose deeds declare their faithfulness,
> And hearty love unto their friend.
> The Gentle Craft in midst of strife
> Yields comfort to a careful life.
>
> A prince by birth I am indeed,
> The which for love forsook this land;
> And when I was in extreme need
> I took the Gentle Craft in hand.
> And by the Gentle Craft alone,
> Long time I lived being still unknown.
>
> Our shoes we sewed with merry notes,
> And by our mirth expelled all moan....'

And so the song runs for another six verses, all having much of the sort of freshness and joviality we find in *The Shoemaker's Holiday*. The narrative continues: 'When the journeymen shoemakers had heard this song, and the fair title that Sir Hugh had given their trade, they engraved the same so deeply in their minds that to this day it could never be razed out: like a remembrance in a marble stone which continueth time out of mind.' A dreadful martyrdom awaits the saintly lovers: Winifred is bled to death and Hugh is required to drink her blood mingled with poison. He does it with a good heart and addresses his last words to the shoemakers:

I drink to you all (quoth he), but I cannot spare you one drop to pledge me. Had I any good thing to give, you should soon receive it. But myself the tyrant doth take, and my flesh is bequeathed to the fowls, so that nothing is left but only my bones to pleasure you withal. And those, if they will do you any good, take them. And so I humbly take my leave, bidding you all farewell.

Some time afterwards, these words are remembered by some shoemakers who are passing the gibbet on which his body was exhibited. They decide to steal the bones by night: 'and because we will turn them to profit and avoid suspicion, we will make divers of our tools with them, and then if any virtue do follow them, the better we shall find it'. So this is what they did, 'and never after did they travel without these tools on their backs: which ever since were called St Hugh's bones'.

In Deloney's second tale we learn the origin of the phrase 'prince am I none, yet am I princely born' which is such a favourite with Simon Eyre. It concerns the French wars in which the hero Crispianus fought for the Gauls against the Persian general, Iphicratis, who had invaded France. Apparently this Iphicratis was a shoemaker's son, and in the exchange of insults which generally preceded a battle, the French taunt him with his lowly origin. 'Thou shalt understand', he replies, 'that a shoemaker's son is a prince born; his fortune made him so, and thou shalt find no less.' It causes some embarrassment amongst the French later on when they find their own champion is also a shoemaker's son. Manly battles are fought amid much courtesy, and Crispianus wins: 'Thus a shoemaker's son was by a shoemaker foiled.' Iphicratis feels that this makes everything all right and they all become good friends.

It is also from Deloney's book that Dekker learnt about Simon Eyre. He follows many of the details closely and most of the alterations are insignificant. For example, it is a Frenchman, not a Dutch skipper, who offers Eyre the goods in his ship: Dekker changed the nationality to bring it in line with

the other part of his story in which Lacy is disguised as the
Dutchman Hans. The one remarkable difference between
Deloney and Dekker is that the means by which Eyre obtains
these goods much more clearly involves some scheming and
sharp-practice in Deloney's account than in Dekker's. In the
original narrative, both Eyre and his wife are quite consciously
ambitious, and even at this stage Eyre is (as he says) 'studying
how to make myself Lord Mayor and thee a Lady'. It is
Mistress Eyre who takes the initiative. There is no need, as
she sees it, to give the Frenchman more than an 'earnest'
—a deposit—(six angels will do) and to say that the bargain
is being conducted on behalf of an alderman. Eyre is to visit
him in the morning, in his normal shoemaker's dress, and
present this pledge. Then in the afternoon, dressed up to the
nines, he will go again, this time as the alderman, exercising
great 'outward courtesy', making the most of the language
barrier, exchange a bill of hand and a credit note, and 'so
come home'. This works well. Eyre sells the goods at a high
price and makes plenty of profit which he invests so as to
make more. It has all been done on false credit and deception,
but nobody appears to be very worried about that: except
that Dekker plays it down, and presents the Eyres rather as
having fortune thrust upon them than consciously scheming
to obtain it.

Most of the details in Dekker's play come from the story
as Deloney tells it. We remember, for instance, that in *The
Shoemaker's Holiday* Eyre feasts the apprentices of London on
Shrove Tuesday because of a vow he had made. The origin is
explained in *The Gentle Craft*. When Simon was a youth, and
the youngest prentice in his shop, he

was often sent to the Conduit for water, where in short time he fell
acquainted with many other prentices coming thither for the same
intent. Now their custom was so, that every Sunday morning

divers of these prentices did use to go to a place near the Conduit to break their fast with pudding-pies, and often they would take Simon along with them. But upon a short time it so fell out that, when he should draw money to pay the shot with the rest, that he had none, whereupon he merrily said to them: 'My faithful friends and Conduit companions, treasurers of the water tankard and main pillars of the pudding house, I may now compare my purse to a barren doe, that yields the keeper no more good than an empty carcass, or to a bad nut which being opened hath never a kernel. Therefore, if it will please you to pardon me at this time and excuse me my part of the shot, I do here vow unto you that if ever I come to be Lord Mayor of this City, I will give a breakfast to all the prentices in London.'

Apparently they were satisfied by this and, although the odds were much against it, Eyre lived to fulfil his vow. 'Such a multitude' was feasted

that besides the Great Hall, all the gardens were set with tables... drums and trumpets were pleasantly sounding....Then after this, Sir Simon Eyre built Leaden Hall appointing that in the midst thereof there should be a market place kept every Monday for leather where the shoemakers of London for their more ease might buy of the tanners without seeking any further. And in the end this worthy man ended his life in London with great honour.

Deloney begins his story of Eyre with a reference to 'our English chronicles' which make mention of him. Of the historical Eyre, we ourselves learn virtually all we know from the famous *Survey of London* which John Stow published first in 1598. Here we read:

In the year next following (1444), the parson and parish of St Dunston, in the east of London, seeing the famous and mighty man (for the words be in the grant, *cum nobilis et potens vir*), Simon Eyre, citizen of London, among other his works of piety, effectually determined to erect and build a certain granary upon the soil of the same city at Leaden hall, of his own charges, for the common utility of the said city, to the amplifying and enlarging of the said granary, granted to Henry Frowicke, then mayor, the aldermen and commonalty, and their successors for ever, all their tenements, with the appurtenances...for the annual rent of four pounds.

In the chapel he had written an inscription which Stow translates thus:

This honourable and famous merchant, Simon Eyre, founder of this work, once mayor of this city, citizen and draper of the same, departed out of this life, the 18th day of September, the year from the Incarnation of Christ 1459, and the 38th year of the reign of King Henry VI.

(Stow's *Survey of London*, Everyman ed., pp. 138–9)

Stow gives details of other of Eyre's benefactions, but there is nothing here more directly relating to the stories told by Deloney.

SELECT BIBLIOGRAPHY

Editions

The Dramatic Works of Thomas Dekker, 4 volumes, ed. F. Bowers (Cambridge, 1953).

The Non-dramatic Works of Thomas Dekker, 5 volumes, ed. A. Grosart (London, 1884–6).

Critical Works

D. J. Enright, Elizabethan and Jacobean Comedy (Pelican Guide to English Literature, vol. 2, 1955).

K. L. Gregg, Thomas Dekker, a Study in Economic and Social Backgrounds (Seattle, 1924).

L. C. Knights, Drama and Society in the Age of Jonson (London, 1937).

General Works

M. St Clare Byrne, Elizabethan Life in Town and Country (London, 1925).

W. Carew Hazlitt, The Livery Companies of the City of London (London, 1892).

Works of Thomas Deloney, ed. F. O. Mann (Oxford, 1912).

Tudor and Stuart Glossary, Skeat and Mayhew (Oxford, 1914).

Stow's Survey of London (Everyman ed., 1912).